Newbold's Biometric Dictionary

For Military and Industry

2nd Edition

Richard D. Newbold, JD, MBA, CIPP/G

authorHOUSE®

AuthorHouse™
1663 Liberty Drive, Suite 200
Bloomington, IN 47403
www.authorhouse.com
Phone: 1-800-839-8640

First published by AuthorHouse 5/27/2008

ISBN: 978-1-4343-4018-4 (sc)

Printed in the United States of America
Bloomington, Indiana

This book is printed on acid-free paper.

Acknowledgments

The author would like to thank the Department of Defense biometrics, counterterrorism, and privacy communities for a wonderful career and a supportive environment. I hope this edition is helpful, and subsequent editions promise to be even more of a comprehensive aid. A special thanks to my wife for her support in my professional endeavors. Also thank you to my editor, who logged many hours and has already signed up for the third edition. And thanks to the end users. I hope that your interest in biometrics will continue to grow as a result of this project. The author is part of a collaborative effort, and I would encourage the reader to visit the Identity Education Consortium (IEC) (www.identityeducation.org). This book is dedicated to our daughter who slipped away to heaven after a long battle with leukemia and is truly missed (www.chloesmiracle.com).

3GPP - 3rd Generation Partnership Project

10-print card - A paper form used to collect both an individual's personal and demographic information along with flat and rolled ink impression fingerprints images. Mainly used in conjunction with an Automated Fingerprint Identification System (AFIS).

10-print match/identification - An absolute positive identification of an individual by corresponding each of his or her 10 fingerprints to those in a system of record, usually performed by an AFIS system and verified by a human fingerprint examiner.

A

AAA - Army Audit Agency

AAWO - Army Asymmetric Warfare Office

ABIS - Automated Biometric Identification System - Department of Defense (DOD) system implemented to improve the U.S. government's ability to track and identify national security threats. The initial focus was on fingerprints gathered by coalition forces from "red force" personnel like detainees, internees, enemy prisoners of war and foreign persons of interest viewed as national security threats. They can be compared with data maintained by the FBI's integrated automated fingerprint identification system (IAFIS), an electronic, searchable database with the fingerprints of approximately 48 million people who have been arrested in the U.S. databases of U.S. government agencies will also eventually be linked so that red force biometric data is searched against multiple databases for any possible matches.

ABIS TM CCB - Automated Biometric Identification System Transaction Manager Configuration Control Board

ACBS - Access Control Badging System

ACCM - Alternative Compensative Control Measures

ACDA - U.S Arms Control and Disarmament Agency

ACM - Anti-Coalition Militia

ACO - Administrative Contracting Officer

ACO (also) - Access Card Office

ACSI - Assistant Chief of Staff for Intelligence - *See also DCSINT.*

ACTD - Advanced Concept Technology Demonstration

ADP - Automated Data Processing

AEA - Atomic Energy Act of 1954 as amended, 42 U.S.C. 2011

AECA - Arms Export Control Act, 22 U.S.C. 2751 *et. seq.*

AFCA - Air Force Communications Agency

AFCAF - Air Force Central Adjudication Facility

AFCEA - Armed Forces Communications and Electronics Association

AFI - Air Force Instruction

AFIS - Automated Fingerprint Identification System - 1) A highly specialized biometric system that compares a single finger image with a database of finger images. AFIS is predominantly used for law enforcement, but is also being put to use in civil applications. In law enforcement, finger images are typically collected from crime scenes (these fingerprints are known as "latents") or directly from criminal suspects after they are arrested. In civilian applications, finger images may be captured by placing a finger on a scanner or by electronically scanning inked impressions on paper. 2) A system originally developed for use by law enforcement agencies, which compares a single fingerprint with a database of fingerprint images. Subsequent developments have seen its use in commercial applications, where a client or customer has their finger image compared with existing personal data by placing a finger on a scanner, or by the scanning of inked paper impressions. 3) A highly specialized biometric system that compares a submitted fingerprint record (usually of multiple fingers) to a database of records, to determine the identity of an individual. AFIS is predominantly used for law enforcement, but is also being used for civil applications (e.g., background checks for soccer coaches, etc). *See also IAFIS.*

AFIS (also) - American Forces Information Service

AFOSF - Air Force Office of Security Forces

AFOSI - Air Force Office of Special Investigations

AFOSP - Air Force Office of Security Police

AFPD - Air Force Policy Directive

AFRL - Air Force Research Laboratory

AFRSSIR - Armed Forces Repository of Specimen Samples for the Identification of Remains

AIA - Air Intelligence Agency

AIA (also) - Army Intelligence Agency

AID - Application Identifier - A globally unique identifier of a card application.

AIMS - Automated Identification Management System

AIS - Automated Information System

AISS - Automated Information Systems Security

AISSP - Automated Information Systems Security Plan

AJCC - Alternate Joint Communications Center - Site R is the Alternate Joint Communications Center (AJCC) located in Raven Rock mountain just over the Pennsylvania State Line near Waynesboro, PA. The facility functions as the disaster recovery site for the JSSC's GMC and DISA GCC. The various service (Army, Navy and Air Force) Emergency Operations Centers (AFEOC) are also

located at Site R. Support is provided 24 hours per day,
7 days per week. The facility's Operations Center, DCS
Technical Control Facility, the Northeast Dial Service
Assistance Center and Information Center provide
planning, installation, operation, and maintenance of over
38 communications systems (switching, transmission, data
distribution, visual information, and power generation)
that support the various customers of the Alternate Joint
Communications Center Site R. The initial concept for
the Alternate Joint Communications Center originated in
1948. After the Soviet Union detonated its first nuclear
weapon in 1949, a high priority was established for
the Joint Command Post to be placed in a protected
location with close proximity to Washington, D.C. for swift
relocation of the National Command Authorities and the
Joint Communications Service. In 1950, President Truman
approved Raven Rock Mountain, Pennsylvania, as part of
Camp Albert Ritchie, Maryland. This new site was named
the Alternate Joint Communications Center (AJCC) Site
R (R for Raven Rock). In 1951, construction of the facility
began, and in 1953, the AJCC Site R became operational.

AKM - Army Knowledge Management

AM - Activity Model

AMC - Army Materiel Command - AMC is the Army's
principal materiel developer. Headquartered in Alexandria,
VA, AMC accomplishes its mission through 11 major
subordinate commands (MSCs) that direct the activities
of numerous depots, arsenals, ammunition plants,
laboratories, test activities, and procurement operations.
AMC is in about 285 locations worldwide, covering more

than 42 states and a dozen foreign countries. Manning these organizations is a work force of more than 65,000 employees, both military and civilian, many with highly developed specialties in weapons development and logistics. Although AMC is over 95-percent civilian, thousands of US Army Reserve (USAR) and Army National Guard (ARNG) soldiers train at AMC installations every year. Essential to the success of AMC training efforts is the highly skilled and experienced civilian work force. From warehouses to production lines, from roadways to railways, Reservists test their combat service support (CSS) skills while completing real-world missions for AMC. AMC's mission is complex and ranges from developing sophisticated weapon systems, to advanced research in such areas as lasers, to maintaining and distributing spare parts. This mission is best summarized by AMC's three core competencies: acquisition excellence, logistics power projection, and technology generation and application. To develop, buy, and maintain materiel for the Army, AMC works closely with industry, colleges and universities, the sister services, and other government agencies to ensure state-of-the-art technology and support are exploited to defend the Nation.

ANACI - Access National Agency Check with written Inquiries

ANSI - American National Standards Institute - A private, non-profit organization that administers and coordinates the U.S. voluntary standardization and conformity assessment system. The mission of ANSI is to enhance both the global competitiveness of U.S. business and the U.S. quality of life.

APB - Acquisition Program Baseline

API - Advance Passenger Information - A set of techniques (including APIS and APP) to declare information on passengers prior to their arrival in a country. Key data in the Advance Passenger Information System (APIS) includes: name, date of birth, gender, document number (such as passport), issuing country, date of expiry, passenger's point of embarkation, passenger's final destination, flight details, and various optional fields.

API (also) - Application Program/Programming Interface - A set of services or instructions used to standardize an application. An API is computer code used by an application developer. Any biometric system that is compatible with the API can be added or interchanged by the application developer. APIs are often described by the degree to which they are high level or low level. High-level means that the interface is close to the application; low-level means that the interface is close to the device.

API (also) - Application Protocol Interface

APIS - Advance Passenger Information System - Developed by the legacy U.S. Customs Service in 1988. Airline carriers collect passenger and crew data from the plane manifest and transmit this data to the CBP Data Center. CBP uses APIS to identify suspect passengers, while facilitating the majority of law-abiding passengers through the clearance process. APIS information can be used to interdict and apprehend potential terrorists before they depart the United States. A structured message format transmitted to the government upon flight closure, usually within 15 to 30 minutes. For the major scheduled carriers, this will be done by the departure control system checking in the passengers onto the flight. This is inherently a batch process done post-flight departure.

APP - Advance Passenger Processing - The standard for flight document data declaration in real-time, This takes place as a real-time interactive transaction during check-in, allowing governments to check against databases to allow or deny passenger check-in.

ARCIC - Army Capabilities Integration Center

AROC - Army Requirements Oversight Council

ASA - Assistant Secretary of the Army

ASA(ALT) - Assistant Secretary of the Army for Acquisition, Logistics, and Technology

ASARC - Army Systems Acquisition Review Council

AS&C - (Office of) Advanced Systems and Concepts - Offices include: 1) CTO - The Comparative Testing Office, which administers the Defense Acquisition Challenge (DAC) and Foreign Comparative Testing (FCT) programs; 2) DAC - solicits challenges to existing technologies to provide companies, individuals, and Defense acquisition programs an on-ramp for increased introduction of innovative and cost-saving technologies; 3) FCT - facilitates the test and evaluation of foreign non-developmental equipment and technology to satisfy U.S. military requirements; 4) OTT - The Office of Technology Transition - which formulates policies and establishes and manages programs that affordably transition advanced technologies from R&D to weapons systems, OTT also assists in the commercialization of defense technologies. Programs include: a) Technology Transfer (including TechLink and TechMatch programs); b) IR&D - Independent Research and Development; c) SBIR - Small Business Innovation Research; d) NATIBO - North American Technology and Industrial Base Organization; e) ManTech - Manufacturing Technology; f) Defense Production Act Title III; and g) Technology Transition Initiative; and 5) JCOS - The Joint and Coalition Operations Support Office, which oversees all activities directly associated with the Joint Warfighting Combatant Commanders (CoComs) and Coalition partners. Among other duties, JCOS also formulates strategic goals and develops architecture for technology demonstration and transition in achieving DoD's transformational goals; and 6) PR&I - Program Resources and Integration Office, which manages funding of all AS&C projects.

ASD(C3I) - Assistant Secretary of Defense (Command, Control, Communications, and Intelligence)

ASD(NII) - Assistant Secretary of Defense (Networks and Information Integration) - The principal OSD staff assistant for the development, oversight, and integration of DoD policies and programs relating to the strategy of information superiority for the Department of Defense. ASD(NII) functions include information policy and information management, command and control, communications, counterintelligence, security, information assurance, information operations, space systems and space policy, intelligence, surveillance and reconnaissance, and intelligence-related activities conducted by the Department. In addition, the ASD(NII) serves as the Chief Information Officer of the Department.

ASD(NII)/DoD CIO - Assistant Secretary of Defense for Networks and Information Integration/DoD Chief Information Officer

ASD(P&L) - Assistant Secretary of Defense (Production and Logistics) - Provides the DUSD(TWP) advice on the production, procurement, deployment, and support of the PSE programs and provides senior-level representation on the PSESG and representation on the PSEAG.

ASD(SO/LIC) - Assistant Secretary of Defense (Special Operations and Low-Intensity Conflict) - Provides the DUSD(TWP) advice on special PSE requirements to support antiterrorist programs, coordinates special equipment requirements for physical security with the PSEAG, provides senior-level representation on the PSESG and representation on the PSEAG.

ASN.1 - Abstract Syntax Notation

ASIC - Application Specific Integrated Circuit - An integrated circuit (silicon chip) that is specially produced for a biometric system to improve performance.

ASIS - American Society for Industrial Security

ASN.1 - Abstract Syntax Notation - A formal language for abstractly describing messages to be exchanged among an extensive range of applications involving the Internet, intelligent network, cellular phones, ground-to-air communications, electronic commerce, secure electronic services, interactive television, intelligent transportation systems, Voice Over IP, and others.

AT - Anti-Terrorism/Antiterrorism

AT&L - Acquisition, Technology, and Logistics

ATA - Antiterrorism Assistance Program - Implemented by the State Department's Bureau of Diplomatic Security, provides foreign partners with training on the identification of fraudulent travel documents. This course is administered jointly with the TSA and Immigration and Customs Enforcement (ICE). ATA trains foreign officials in preventing unauthorized access to aircraft. In addition, ATA develops regional law enforcement relationships and mechanisms for sharing information related to terrorist threats and operations.

ATEC - Army Test and Evaluation Command

AT/FP - Anti-Terrorism/Force Protection

ATIS - Alliance for Telecommunications Industry Solutions

ATL or AT&L - Acquisition, Technology, and Logistics

ATSD(AE) - Assistant to the Secretary of Defense (Atomic Energy) - Serves as the advisor and focal point for issues on nuclear weapons security and provides senior-level representation on the PSESG and representation on the PSEAG.

AVIDS - Automated Verification Identification System

AWIPT - Architecture Working Integrated Product Team

Acceptability - Indicates the degree of approval of a technology by the public in everyday life.

Access control - Process of granting access to information system resources only to authorized users, programs, processes, or other systems.

Accuracy - A catch-all phrase for describing how well a biometric system performs. The actual statistic for performance will vary by task (verification, open-set identification, and closed-set identification). *See also d prime, detection error trade-off (DET), detect and identification rate, equal error rate, false acceptance rate (FAR), false alarm rate (FAR), false match rate, false non-match rate, false reject rate, identification rate, performance, and verification rate.*

Acoustic emission - A proprietary technique used in signature verification. As a user writes on a paper surface, the movement of the pen tip over the paper fibers generates acoustic emissions that are transmitted in the form of stress waves within the material of a writing block beneath the document being signed. The structure-borne elastic waves behave in materials in a similar way to sound waves in air and can be detected by a sensor attached to the writing block.

Acquisition device - The hardware used to acquire biometric samples. The following acquisition devices are associated with each biometric technology.

Active attack - When an impostor intentionally modifies, simulates or reproduces a biometric sample and submits this sample into the system in an attempt to initiate a false match.

Active imposter acceptance - Acceptance of a biometric sample submitted by someone attempting to gain illegal entry to a biometric system.

Adaptation - The process of automatically updating or refreshing a reference template.

Adaptation (model/template adaptation) - The use of a newly captured and verified biometric sample to automatically update or refresh the enrolled reference template. Used to minimize the effects of template aging.

Agency record - The products of data compilation, such as all books, papers, maps, and photographs, machine readable materials, inclusive of those in electronic form or format, or other documentary materials, regardless of physical form or characteristics, made or received by an agency of the United States Government under Federal law in connection with the transaction of public business and in Department of Defense possession and control at the time the FOIA request is made.

Algorithm - 1) A sequence of instructions that tell a biometric system how to solve a particular problem. An algorithm will have a finite number of steps and is typically used by the biometric engine to compute whether a biometric sample and template is a match. 2) A limited sequence of instructions or steps that tells a computer system how to solve a particular problem. A biometric system will have multiple algorithms, for example: image processing, template generation, comparisons, etc. 3) A sequence of instructions that tells a system how to solve a problem. Used by biometric systems, for example: to tell whether a sample and a template are a match. Cryptographic algorithms are used to encrypt sensitive data files, to encrypt and decrypt messages, and to digitally sign documents. 4) A step-by-step procedure for carrying out a mathematical computation or a transformation of data, usually used in reference to work performed by a computer.

Algorithm identifier - A PIV algorithm identifier is a one-byte identifier that specifies a cryptographic algorithm and key size. For symmetric cryptographic operations, the algorithm identifier also specifies a mode of operation (i.e., CBC or ECB).

Application - 1) A hardware/software system implemented to satisfy a set of requirements. In this context, an application incorporates a biometric system to satisfy a subset of requirements related to the verification or identification of an end user's identity so that the end user's identifier can be used to facilitate the end user's interaction with the system. 2) Software program that performs a specific function directly for a user and can be executed without access to system control, monitoring, or administrative privileges. Examples include office automation, electronic mail, web services, and major functional or mission software programs.

Application developer - An individual entrusted with developing and implementing a biometric application.

Application profile - Conforming subsets or combinations of base standards used to provide specific functions. Application profiles identify the use of particular options available in base standards, and provide a basis for the interchange of data between applications and interoperability of systems.

Application session - The period of time within a card session between when a card application is selected and a different card application is selected or the card session ends.

Arch - A fingerprint pattern in which the friction ridges enter from one side, make a rise in the center and exit on the opposite side. The pattern will contain no true delta point. *See also delta point, loop, and whorl.*

Asymmetric cryptography - Asymmetric means that two parts of a thing are not similar (not symmetric). In asymmetric cryptography a private key is used for creating a digital signature, and the related public key is used for verifying the signature. Because the keys for each process are different the processes are described as being asymmetric. Asymmetric cryptography is a synonym for public key cryptography.

Asymmetric key - In asymmetric cryptography, one key of a pair of asymmetric keys (a public key and a private key). *See public key cryptography.*

Asymmetric warfare - A term that describes a military situation in which two belligerents of unequal strength interact and take advantage of their respective strengths and weaknesses. This interaction often involves strategies and tactics outside the bounds of conventional warfare.

Asynchronous multimodality - Systems that require that a user verify through more than one biometric in sequence. Asynchronous multimodal solutions are comprised of one, two, or three distinct authentication processes. A typical user interaction will consist of verification on finger scan, then face if finger is successful.

ATOMAL - NATO Marking for U.S./UK Atomic Information

Attempt - The submission of a single set of biometric sample to a biometric system for identification or verification. Some biometric systems permit more than one attempt to identify or verify an individual. *See also biometric sample, identification, and verification.*

Attended versus Non-attended - A fourth partition is "attended/unattended," and refers to whether the use of the biometric device during operation will be observed and guided by system management. Non-cooperative applications will generally require supervised operation, while cooperative operation may or may not. Nearly all systems supervise the enrollment process, although some do not.

Attribute authority - An entity, recognized by a Certificate Management Authority, as having the authority to verify the association of attributes to an identity.

Audit trail - In computer/network systems, a record of events (protocols, written documents, and other evidence), which can be used to trace the activities and usage of a system. Such material is crucial when tracking down successful attacks/attackers, determining how the attacks happened, and being able to use this evidence in a court of law.

Authenticatable entity - An entity that can successfully participate in an authentication protocol with a card application.

Authentication - 1) The process of establishing confidence in the truth of some claim. The claim could be any declarative statement. For example: "This individual's name is 'Joseph K.'" or "This child is more than five feet tall." In biometrics, "authentication" is sometimes used as a generic synonym for verification. 2) The process of establishing the validity of the user attempting to gain access to a system. Primary authentication methods are access; passwords (something the user knows); access tokens (something the user owns); biometrics; geography (a workstation, for example); security measures designed to establish the validity of a transmission, message, or originator, or a means of verifying an individual's authorization to receive specific categories of information. 2) The process of determining whether someone or something is actually who or what it asserts itself to be. In the context of an access control system, it refers to the process of identifying the individual or system requesting access, by checking the credentials presented against the information stored in the system. 3) Refers to either biometric process of verification or identification of an individual. 4) The use of a biometric characteristic for the purpose verification or identification of an individual.

5) The process of determining an individual's identity, either by verification or by identification. 6) The process of establishing that an individual, previously identified and with whom a business relationship has been established, is the same as the individual who initially created the relationship. This is generally done by presenting information that is known only to the individual and the organization. Authentication is also a security measure designed to establish the validity of a transmission, message, or originator, or a means of verifying an individual's authorization to receive specific categories of information. *See also identify and verification.*

Authentication assurance level - *See identity authentication assurance level.*

Authentication routine - A cryptographic process used to validate a user, card, terminal, or message contents. Also known as a handshake, the routine uses important data to create a code that can be verified in real time or batch mode. *See also verification.*

Authentication system - A system comprising a verification device and a function that determines whether the individual sitting for the attempt is an impostor or the individual of correct identity, based on the results from the verification device.

Authorization - 1) The assignment of access rights to an individual or system in an identity management system or access control system (*see also user provisioning*). 2) The decision process by an access control system of determining whether or not access should be granted at the time an access request is made based upon the rights that have been assigned.

Auto-correlation - A proprietary finger scanning technique. Two identical finger images are overlaid in the auto-correlation process, so that light and dark areas, known as Moiré fringes, are created.

Automatic ID/Auto ID - An umbrella term for any biometric system or other security technology that uses automatic means to check identity, this applies to both one-to-one verification and one-to-many identification.

B

B - Number of binning and filtering methods.

BAH - Booz Allen Hamilton

BAT - Biometric Automated Toolset

BCC - Border Crossing Credentials

BCCB - Biometrics Configuration Control Board

BCE - Baseline Cost Estimate

BDI - Ballistic Defense Initiative

BDST - Biometrics Decision Support Tool - The BDST is still in the inception phase and is in co-development with West Virginia University and the Biometrics Fusion Center. The concept is refined and endorsed as an NSTC Biometrics Working Group ICP.

BEM - Biometric Evaluation Methodology

BER - Basic Encoding Rules

BES - Budget Estimate Submission

BFC - Biometrics Fusion Center - Located in Clarksburg, WV, BFC is a biometrics test and evaluation facility.

BFT - Basic Functional Testing - The purpose of the BFT is to investigate product functionality and to document the findings. These findings are published in the DBEKS Reports section of the site to assist potential users in making decisions about purchasing biometric devices.

BHSUG - Biometrics in Human Services User Group

BI - Background Investigation - Conducted by DIS and is much more extensive than a NAC. It is designed to develop information as to whether the access to classified information by the person being investigated is clearly consistent with the interests of national security. It includes an NAC and probes deeply into the loyalty, integrity, and reputation of the individual.

BioAPI - Biometrics Application Programming Interface - Defines the application programming interface and service provider interface for a standard biometric technology interface. BioAPI V1.0, developed by the BioAPI consortium, and released in March 2000. Designed to produce a standard biometric API aiding developers and consumers, The BioAPI enables biometric devices to be easily installed, integrated or swapped within the overall system architecture.

BIR - Biometric Identification Record - Includes the reference template and other data associated with the user.

BIR (also) - Biometric Intelligence Repository

BISA - Biometric Identification System for Access - BISA is a $75 million DoD project that will assist in force protection initiatives for U.S. installations in Iraq. In order for local foreign-nationals and other non-U.S. citizens who wish to work on U.S. bases to be granted access, an improved security system is necessary to ensure identity. BISA involves collecting facial, fingerprint and iris data from employees and placing the data on a smart card. To gain admittance, employees simply scan their card and place their finger on wireless biometric readers at entry stations. The need to improve security at overseas installations originates from a memo sent by then-Deputy Secretary of Defense Paul Wolfowitz in May 2005.

BIT - Biometric Information Template - BIT, which contains the Biometric Header Template BHT, the biometric data Block BDB consisting of Biometric reference data possibly followed by a payload and optional DOs related to security.

BITS - Biometrics Implementation Tracking System - The BITS is a tool that is an authoritative system in which biometric implementations/projects may be tracked and reported. The tool will also serve as a repository of information pertaining to biometric pilots, as a historical record of the status and progress of DoD biometric implementations.

BMO - Biometrics Management Office - *See BTF.*

BOB - Biometric Operations Board

BOD - Board of Directors

BPAC - Budget Program Activity Code

BPL - Biometrics Products List

BPO - Biometrics Project Office

BQCP - Biometric Quick Capture Platform - Designed to implement a new biometric (fingerprint) system that includes a quick capture live-scan fingerprint acquisition device with required software, a laptop, and a portable satellite unit. This capability will allow for a real-time electronic connection between U.S. deployed entities and the FBI to enable a search of the FBI's entire fingerprint database.

BSP - Biometric Service Provider - A BSP typically performs biometric operations like enrollment, verification, and identification. They control biometric hardware devices, implement any custom or standard biometric algorithms for processing and matching, and support any custom or standard biometric data formats.

BSWG - Biometric Standards Working Group

BTF - Biometrics Task Force - The Biometrics Task Force leads Department of Defense activities to program, integrate, and synchronize biometric technologies and capabilities and to operate and maintain DoD's authoritative biometric database to support the National Security Strategy.

BTT - Biometrics Tiger Team

Base standard - Fundamental and generalized procedures, provides an infrastructure that can be used by a variety of applications, each of which can make its own selection from the options offered by them.

Battlespace digitization - Designed to improve military operational effectiveness by integrating weapons platforms, sensor networks, UC2, intelligence, and network-centric operations. This military doctrine reflects that in the future, military operations will be merged into joint operations rather than take place in separate battlespaces under the domain of individual armed services.

Behavioral biometric - Characterized by a behavioral trait that is learned and acquired over time rather than a physiological characteristic. Examples include signature, keystroke dynamics, speech, etc.

Behavioral biometric characteristic - A biometric characteristic that is learned and acquired over time rather than one based primarily on biology. All biometric characteristics depend somewhat upon both behavioral and biological characteristic. Examples of biometric modalities for which behavioral characteristics may dominate include signature recognition and keystroke dynamics. *See also biological biometric characteristic.*

Benchmarking - The process of comparing measured performance against a standard, openly available, reference.

Bifurcation - The point in a fingerprint where a friction ridge divides or splits to form two ridges that continue past the point of division for a distance that is at least equal to the spacing between adjacent ridges at the point of bifurcation. *See also friction ridge, minutia(e) point, and ridge ending.*

Binning - Process of parsing (examining) or classifying data in order to accelerate and/or improve biometric matching. This allows a database of biometric data to be pre-sorted in order to speed up the process of matching captured biometric data with comparison data.

Binning error - Occurs if the enrollment template and a subsequent sample from the same biometric feature on the same user are placed in different partitions.

BioAPI Consortium - Founded to develop a biometric Application Programming Interface (API) that brings platform and device independence to application programmers and biometric service providers. The BioAPI Consortium is a group of over 120 companies and organizations that have a common interest in promoting the growth of the biometrics market. The BioAPI Consortium developed a specification and reference implementation for a standardized API that is compatible with a wide range of biometric application programs and a broad spectrum of biometric technologies.

BioAPI specification - Defines the application programming interface and service provider interface for a standard biometric technology interface.

Biological biometric characteristic - A biometric characteristic based primarily on an anatomical or physiological characteristic, rather than a learned behavior. All biometric characteristics depend somewhat upon both behavioral and biological characteristic. Examples of biometric modalities for which biological characteristics may dominate include fingerprint and hand geometry. *See also behavioral biometric characteristic.*

Biometric (adjective) - Of or pertaining to technologies that utilize behavioral or physiological characteristics to determine or verify identity. For example, "Do you plan to use biometric identification or older types of identification?"

Biometric (noun) - 1) Measurable physical characteristic or personal behavioral trait used to recognize the identity or verify the claimed identity of an individual. 2) A measurable physiological or behavioral characteristic which can be used to reliably distinguish one individual from another. 3) One of various technologies that utilize behavioral or physiological characteristics to determine or verify identity. For example, "Fingerprint is a commonly used biometric." Plural form also acceptable: "Retina-scan and iris-scan are eye-based biometrics." 4) Field relating to biometric identification. For example, "What is the future of biometrics?"

Biometric authentication mode - The way biometric data (e.g., fingerprints) is used for authentication. The mode chosen for a biometric installation depends on the specific needs of a site, where either convenience or security may be emphasized. BioCert fingerprint devices may use either of two biometric authentication modes, identification or verification.

Biometric characteristic - A physical characteristic or a behavioral characteristic.

Biometric Consortium - An open forum to share information throughout government, industry, and academia.

Biometric data - 1) A catch-all phrase for computer data created during a biometric process. It encompasses raw sensor observations, biometric samples, models, templates and/or similarity scores. 2) The extracted information taken from the biometric sample and used to either build a reference template or to compare against a previously created reference template. 3) The information extracted from the biometric sample and used either to build a reference template (template data) or to compare against a previously created reference template (comparison data).

Biometric device - The logical component of a biometric system responsible for the capture of a biometric characteristic.

Biometric engine - 1) The software element of the biometric system which processes biometric data during the stages of enrollment and capture, extraction, comparison and matching. 2) The aspect of a biometric security system that administers the biometric data during the enrollment, capture, extraction, comparison, and matching stages of the process. The biometric engine is a software program that works in conjunction with the hardware devices that a biometric system uses.

Biometric evidence - A measurable, physical characteristic or personal behavioral trait available on any physical support. For example, latent fingerprint, DNA of one hair, record of a voice, signature on a document, etc.

Biometric feature - A representation from a biometric sample extracted by the extraction system.

Biometric fingerprinting - Digital image capture of friction ridges and/or a template from friction ridges.

Biometric glossary - Data is used to describe the information collected during an enrollment, verification, or identification process but does not apply to end user information such as user name, demographic information and authorizations.

Biometric identification record - Any biometric data that is returned to the application, including raw data, intermediate data, or processed sample(s) ready for verification or identification, as well as enrollment data.

Biometric image - A Biometric sample in the form of an image of an end-user biometric captured by a biometric system.

Biometric information - Information needed by the outside world to construct the verification data.

Biometric passport - A biometric passport is a combined paper and electronic identity document that uses biometrics to authenticate the citizenship of travelers. The passport's critical information is stored on a small RFID computer chip, much like information stored on smartcards. Like some smartcards, the passport book design calls for an embedded contactless chip that is able to hold digital signature data to ensure the integrity of the passport and the biometric data.

Biometric performance - A quantifiable measurement of the accuracy and speed, as well as other desirable attributes, of a biometric measurement, algorithm, or system.

Biometric performance metric - A quantifiable assessment of the speed, accuracy, or other characteristic of a biometric algorithm or system.

Biometric properties - Include universality, uniqueness, permanence, collectability, performance, acceptability, and circumvention.

Biometric reference template - A data set which defines a biometric measurement of a person which is used as a basis for comparison against a subsequently submitted biometric sample(s).

Biometric sample - 1) The identifiable, unprocessed image or recording of a physiological or behavioral characteristic, acquired during submission, used to generate biometric templates. Also referred to as biometric data. 2) Information or computer data obtained from a biometric sensor device. Examples are images of a face or fingerprint. 3) Data representing a biometric characteristic of a user as captured by a biometric system; raw data representing a biometric characteristic of an end-user as captured by a biometric system (e.g., the image of a fingerprint).

Biometric system - An automated system capable of 1) Capturing a biometric sample from an end-user; 2) Extracting biometric data from that sample; 3) Comparing the biometric data with that contained in one or more reference templates; 4) Deciding how well they match; and 5) Indicating whether or not an identification or verification of identity has been achieved. A biometric system may be a component of a larger system. Multiple individual components (such as sensor, matching algorithm, and result display) that combine to make a fully operational system.

Biometric taxonomy - A method of classifying biometrics. For example, an application may be classified as: Cooperative vs. Non-Cooperative User/Overt vs. Covert Biometric System/Habituated vs. Non-Habituated User/ Supervised vs. Unsupervised User/Standard Environment vs. Non Standard Environment. Co-operative refers to a willing end user participating in a biometric application. Overt refers to an undisguised and candid use of a biometric system. Habituated means that an end user is familiar with the workings of the biometric system and application, supervised means that trained personnel guide an end user through the biometric application. Standard environment refers to unchanging and non-volatile surroundings and climate.

Biometric technology - A classification of a biometric system by the type of biometric.

Biometric template - *See template.*

Biometric transaction - The sequence of events by which a person attempts to have an identity recognized or verified by a biometric identification device culminating in a person's acceptance or rejection.

Biometric type - A set of associated biometrics subtypes with common properties and characteristics.

Biometric verification - Process of verifying by a one-to-one comparison of the user's verification data against reference data, whether or not the user is the one he/she claims.

Biometric word list - List of words that can be used to authentically and reliably communicate numeric information by voice. The words in the list correspond to one of each of the 256 unique byte values, and are carefully chosen for their phonetic distinctiveness. The properties of the human voice serve as the authentication mechanism.

Biometrics - 1) The automated use of physiological or behavioral characteristics to determine or verify identity. 2) A general term used alternatively to describe a characteristic or a process. As a characteristic: measurable biological (anatomical and physiological) and behavioral characteristic that can be used for automated recognition. As a process: automated methods of recognizing an individual based on measurable biological (anatomical and physiological) and behavioral characteristics.

Biometrics glossary - 1) 10 rolled fingerprints, a minimum of five mug shots from varying angles, and an oral swab to collect DNA. 2) Generic term sometimes used in the biometrics community to discuss a biometric system. *See also AFIS.*

Biometrics or biometrics sub-type - A category of measurable, physical characteristic, or personal behavioral trait of a human being used to authenticate the claimed identity.

Blue Force database - A database containing biometrics and information for known friendly forces.

Body odor - A physical biometric that analyses the unique chemical pattern made up by human body smell.

Booking - The process of capturing inked finger images on paper, for subsequent processing by an AFIS.

Breeder document - A document, such as a birth certificate, that is used by an identification issuer to establish the identity of an applicant.

Buddy-punching - When employees manually punch each other's time cards to falsify their work hours. Biometric systems can end that practice by verifying employees' identities as they clock in and out, most commonly through a fingertip or handprint reader.

Buffer overflow - Most common cause of current security vulnerabilities. A buffer overflow occurs when more data is put into a temporary data storage area (buffer) than the buffer can hold. Because buffers can only hold a finite amount of data, the extra information can overflow into adjacent buffers, corrupting or overwriting the data in them. Programming errors are one of the most frequent causes of buffer overflow problems. In attacks, which exploit buffer vulnerabilities, extra data is sent to the buffer with code designed to trigger specific actions, and which can damage files, change data, or disclose confidential information. Buffer overflow attacks may have arisen from poor use of the C programming language.

C

(C) - Confidential

C3I - Command, Control, Communications and Intelligence

C4 - Command, Control, Communications and Computers

C4ISR - Command, Control, Communications, Computers, Intelligence, Surveillance, and Reconnaissance

CA - Certificate/Certification Authority - An authority trusted by one or more users to create and assign certificates.

CAC - Common Access Card - The standards identification card for active duty personnel (to include the selected reserve), DoD civilian personnel, and eligible contractor personnel. It is the principal card used to enable physical access to buildings and controlled spaces and can be used to gain access to the department's computer networks and systems. The card, which accommodates an integrated circuit chip, also contains other relevant media such as magnetic strips and bar codes.

CAE - Component Acquisition Executive

CAF - Central Adjudication Facility

CAGE - Commercial and Government Entity

CAJIT - Central American Joint Intelligence Team

CAO - Contract Administration Office

CAPI - Crypto Application Program Interface

CBA - Capabilities Based Assessment

CBAB - Central Command Biometrics Advisory Board - Created by the Chief of Staff to assist the Commander in execution of overall management of CENTCOM biometrics activities, the DBAB is co chaired by the J2X and the Deputy Director of the Joint Security Directorate and is comprised of broad membership from across the CENTCOM staff.

CBC - Cipher Block Chaining

CBEFF - Common Biometric Exchange Formats Framework - 1) A standard that provides the ability for a system to identify, and interface with, multiple biometric systems, and to exchange data between system components. 2) Describes a set of data elements necessary to support biometric technologies in a common way. These data can be placed in a single file used to exchange biometric information between different system components or between systems. The result promotes interoperability of biometric-based application programs and systems developed by different vendors by allowing biometric data interchange.

CBO - Congressional Budget Office

CBP - Customs and Border Protection

CBRNE - Chemical Biological Radiological Nuclear High Yield Explosives

CCA - Cross Certification Agreement

CCC - Card Capability Container

CCD - Charge-Coupled Device - A semiconductor device that records images electronically.

CCD (also) - Consolidated Consular Database

CCEM/CEM - Common Criteria Evaluation Methodology

CCF - Central Clearance Facility

CCI - Controlled Cryptographic Item

CCTV - Closed Circuit Television

CDD - Capabilities Development Document

CDRT - Capabilities Development Rapid Transition

CENTCOM - U.S. Central Command

CEP - Continuous Evaluation Program

CER - Crossover Error Rate - 1) A comparison metric for different biometric devices and technologies. 2) The error rate at which FAR equals FRR. The lower the CER, the more accurate and reliable the biometric device.

CERS - Contractor Employee Resources System

CERT - Computer Emergency Response Team

CESG - Communications Electronics Security Group - This organization is conducting a biometrics program for the UK as part of a government modernization initiative.

CFR - Code of Federal Regulations

CHUID - Cardholder Unique Identifier - In FIPS 201, a standard data model for cardholder identification data.

CI - Counter-Intelligence/Counterintelligence - CI is the activity of preventing the enemy from obtaining secret information, such as careful classification and control of sensitive information and spreading disinformation. It encompasses information collection, analysis, investigations and operations conducted to identify and neutralized espionage and foreign intelligence activities, the intelligence-related activities of terrorists, and adversary efforts to degrade, manipulate, or covertly influence U.S. intelligence, political processes, policy, or public opinion.

CI (also) - Critical Information

CIA - Central Intelligence Agency - CIA is an intelligence agency of the United States Government. Its primary function is obtaining and analyzing information about foreign governments, corporations, and persons, and reporting such information to the branches of the Government. Its secondary function is propaganda or public relations, overt and covert information dissemination, both true and false, and influencing others to decide in favor of the United States Government. Its headquarters are in the community of Langley in the McLean CDP of Fairfax County, Virginia, a few miles up the Potomac River from downtown Washington, D.C. The CIA is part of the American Intelligence Community, led by the Director of National Intelligence (DNI). The role and functions of the CIA are roughly equivalent to those of the United Kingdom's Secret Intelligence Service (MI6) and Israel's Mossad.

CID - Criminal Investigation Division - *See CIDC.*

CIDC - Criminal Investigation Division Command

CIFA - Counterintelligence Field Activity

CIK - Crypto-Ignition Key

CINC - Commander in Chief

CINCs - Commander in Chiefs

CIO - Central Imagery Office

CIO (also) - Chief Information Officer

CIP - Critical Infrastructure Protection - CIP is a Presidential directive (PDD-63) that calls for a national effort to assure the security of the increasingly vulnerable and interconnected infrastructures of the United States. In July 1996, President Bill Clinton issued Executive Order Critical Infrastructure Protection. This order stated that certain national infrastructures are critical to the national and economic security of the United States and the well being of its citizenry. The critical infrastructure of the United States is comprised of the systems and networks that are so essential that if one or more is incapacitated or destroyed, an entire region, if not the defense or economic security of the nation, could be debilitated.

CITeR - Center for Identification Technology and Research - CITeR is the first National Science Foundation (NSF) Industry/University Cooperative Research Center (IUCRC) focusing on serving its membership in the rapidly growing area of Biometric Identification Technology.

CJCS - Chairman, Joint Chiefs of Staff

CJCSI - Chairman of the Joint Chiefs of Staff Instruction

CJCSM - Chairman of the Joint Chiefs of Staff Manual

CJIS - Criminal Justice Information Services Division - CJIS is a division of the United States Federal Bureau of Investigation (FBI). A computerized criminal justice information system that is a counterpart of FBI's National Crime Information Center (NCIC) in Washington, and is maintained by Department of Justice (DOJ) in each state. It is available to authorized local, state, and federal law enforcement and criminal justice agencies via any of the three law enforcement communication systems: 1) National Law Enforcement Telecommunications System (NLETS); 2) California Law Enforcement Telecommunications System (CLETS); and the International Law Enforcement Telecommunications System (INLETS). Usually CJIS offers a much wider range of information nationwide and more precise inquiry search parameters than NCIC. CJIS consists of several databases and one subsystem, and its retrieval and update capabilities are online.

CJM3IEM - Combined Joint Multilateral Master Military Information Exchange Memorandum of Understanding

CJMIEA - Combined Joint Military Information Exchange Annex

CLASS - Classified By

CLEAR - Citizen Law Enforcement Analysis and Reporting System

CLO - Cryptographic Network Log-On

CM - Classification Management

CMC - Cumulative Match Characteristic - A method of showing measured accuracy performance of a biometric system operating in the closed-set identification task. Templates are compared and ranked based on their similarity. The CMC shows how often the individual's template appears in the ranks (1, 5, 10, 100, etc.), based on the match rate. A CMC compares the rank (1, 5, 10, 100, etc.) versus identification rate.

CMF - Criminal Master File

CMOS - Complementary Metal Oxide Semiconductor - A type of integrated circuit used by some biometric systems because of its low power consumption.

CMS - Card Management System

CNO - Chief of Naval Operations

CNRI - Corporation for National Research Initiatives

CNWDI - Critical Nuclear Weapon Design Information

COA - Course of Action

COCOM - Combatant Command

CODIS - Combined DNA Index System

COMINT - Communications Intelligence

COMPUSEC - Computer Security

COMSEC - Communications Security - Protective measures taken to deny unauthorized persons information derived from telecommunications of the U.S. Government related to national security and to ensure the authenticity of such communications. Such protection results from the application of security measures (including cryptosecurity, transmission security, emissions security, and jamming resistance) to telecommunications and to electrical systems generating, handling, processing, or using national security or national security-related information. It also includes the application of physical security measures to COMSEC information or materials.

CONOPS - Concept of Operations

CONUS - Continental United States

COOP - Continuity of Operations Plan

COR - Contracting Officer Representative

CORVET - Coordinate Operational Resources for Voice Exploitation and Technology

COSMIC - NATO Top Secret

CoT - Circle of Trust

COTR - Contracting Officer's Technical Representative

COTS - Commercial off-the-shelf - COTS is a term for software or hardware products that are ready-made and available for sale to the general public. They are often used as alternatives to in-house developments or one-off government-funded developments (GOTS). The use of COTS is being mandated across many government and business programs, as they may offer significant savings in procurement and maintenance. The motivation for using COTS components is that they will reduce overall system development costs and involve less development time because the components can be bought instead of being developed from scratch. This could prove to be useful for software development because of the ever-increasing costs. Many considered COTS to be the silver bullet during the nineties but COTS development came with many not so obvious trade-offs. Overall cost and development time can definitely be reduced, but often at the cost of an increase in software component integration work and a dependency on a third-party component vendor.

C-PR - Confidential Periodic Reinvestigation

CPD - Capability Production Document

CPNI - Customer Proprietary Network Information

CPSO - Contractor Program Security Officer

CPU - Central Processing Unit

CRA - Command Resolution Authority

CRADA - Collaborative Research and Development Agreement

CRL - Certificate Revocation List - A list of digital certificates that have been revoked (cancelled) before their expiration date.

CRS - Congressional Research Service

CSA - Cognizant Security Agency

CSA (also) - Chief of Staff, Army

CSLA - Army Communications Security Logistics Activity

CSO - Cognizant Security Office

CSS - Constant Surveillance Service

CSSI - Case Summary Sheet Information

CSSO - Contractor Special Security Officer

CSSR - Case Summary Sheet Recommendation

CSSWG - Contractor SAP/SAR Security Working Group

CT - Counter Terrorism/Counterterrorism

CT (also) - Information and Communications Technology

CTAG - Counterterrorism Action Group - The G8
established CTAG composed of donor Countries, including
the G8 members and others, to expand and coordinate
training and assistance for countries that have the
political will but lack the capacity to combat terror. CTAG
provides an active forum for donor countries to coordinate
counterterrorism cooperation with, and assistance
to, countries in support of the UN Counterterrorism
Committee's efforts to oversee implementation of UN
Security Council Resolution 1373. This resolution obligates
all states to deny safe haven to those who finance, plan,
support, or commit terrorist acts. CTAG has coordinated
efforts to assist countries to assess and improve airport
security and has promoted and assisted with the
implementation of travel security and facilitation standards
and practices developed by G8's SAFTI. In conjunction

with the Asia-Pacific Economic Coordination (APEC), CTAG also has worked to improve APEC countries' port and maritime security.

CTC - Counterterrorist Center - Has its origins in the events in the mid-1980s. A series of high-profile terrorist attacks--including the hijacking of TWA flight 847 which resulted in the murder of a U.S. Navy diver galvanized U.S. policymakers to take the offensive against international terrorism. A task force was formed in 1986, chaired by then Vice President George Bush to address the problem of international terrorism. The task force concluded that U.S. Government agencies collected information on terrorism but did not aggressively operate to disrupt terrorist activities. As a result of these findings, then Director of Central Intelligence, William Casey, created the CTC and directed it to preempt, disrupt, and defeat terrorists.

CTC (also) - Counterterrorism Committee

CTO - Comparative Testing Office

CTTSO - Countering Terrorism Technology Support Office

CTS - COSMIC Top Secret

(CTSA) - COSMIC Top Secret ATOMAL

CUI - Controlled Unclassified Information

CVA - Central Verification Activity (DSS)

CyA - Cyber Assault - Following infiltration, the destruction of software and data in the system, or attack on a system that damages the system capabilities, includes viruses and overload of systems through e-mail (i.e., e-mail overflow).

CyC - Cyber Crime - Cyber attacks without the intent to affect national security or to further operations against national security

CyI - Cyber Infiltration - Penetration of the defenses of a software-controlled system such that the system can be manipulated, assaulted, or raided.

CyM - Cyber Manipulation - Following infiltration, the control of a system via its software, which leaves the system intact, then uses the capabilities of the system to do damage. For example, using an electric utility's software to turn off power.

CyR - Cyber Raid - Following infiltration, the manipulation or acquisition of data within the system, which leaves the system intact, results in transfer, destruction, or alteration of data. For example, stealing e-mail or taking password lists from a mail server.

CyW - Cyber Warfare - Any act intended to compel an opponent to fulfill our national will, executed against the software controlling processes within an opponent's system. CyW includes the following modes of cyber attack: cyber infiltration, cyber manipulation, cyber assault, and cyber raid.

CW - Codeword

Capacitance - A finger image capture technique that senses an electrical charge, from the contact of ridges, when a finger is placed on the surface of a sensor.

Capacitive sensor - Consists of a row/column configuration of tiny metal electrodes. Every column is linked to a pair of sample-and-hold circuits. The fingerprint image is recorded in sequence row by row as each metal electrode acts as one capacitor plate and the contacting finger acts as the second plate. A passivation layer on the surface of the device forms the dielectric between these two plates. Variations in the dielectric between a fingerprint ridge (mainly water) and a valley (air) cause the capacitance to vary locally. Pressing the finger onto the sensor creates varying capacitor values across the arrays that are then converted into an image of the fingerprint. The values of the array are determined by the contour (ridges and valleys) of the fingerprint. The sensor quickly captures several images of the fingerprint and selects the highest quality image.

Capture - The process of taking a biometric sample from the user. In capture, a sample of the user's biometric is acquired using the required sensor (e.g., camera, microphone, fingerprint scanner); the process of collecting a biometric sample from an individual via a sensor. *See also submission.*

Capture - The process of taking a biometric sample via a sensor from a user.

Capturing - Taking a raw biometric sample.

Card application - A set of data objects and card commands that can be selected using an application identifier.

Certificate - A digital representation of information that at least 1) identifies the certification authority issuing it; 2) names or identifies its subscriber; 3) contains the subscriber's public key; 4) identifies its operational period, and 5) is digitally signed by the certification authority issuing it.

Certificate - *See digital certificate.*

Certificate/certification authority - Person or company who issues, revokes and manages digital certificates to subscribers. A CA acts as a trusted "third party" certifying the identity of subscribers to anyone who receives a digitally signed message.

Certificate expiry - The date after which a user's certificate should no longer be trusted. The certificate expiration date is contained within the certificate.

Certificate request - A subscriber receives a digital certificate by requesting one from a certificate authority.

Certificate revocation - A certificate shall be revoked when the binding between the subject and the subject's public key defined within a certificate is no longer considered valid.

Challenge response - A method used to confirm the presence of a person by eliciting direct responses from the individual. Responses can be either voluntary or involuntary. In a voluntary response, the end user will consciously react to something that the system presents. In an involuntary response, the end user's body automatically responds to a stimulus. A challenge response can be used to protect the system against attacks. *See also liveness detection.*

Circumvention - How easy or difficult it is to fool an authentication system.

Claim, defrauder - An individual making a false negative claim on identity (either implicit or explicit).

Claim, explicit negative - An individual claims to not be previously enrolled in a system and claims to not have a system identifier.

Claim, explicit positive - An individual claims to be enrolled and provides an identifier known by the biometric system (this identifier can be used to select the individual's biometric reference template for direct comparison in a verification attempt).

Claim, genuine - An individual making a truthful claim on identity (either implicit or explicit).

Claim, implicit negative - An individual claims to not be previously enrolled in the biometric system.

Claim, implicit positive - An individual claims to be enrolled in the biometric system.

Claim, impostor - An individual making a false positive claim on identity (either implicit or explicit).

Claim of identity - 1) A statement that a person is or is not the source of a reference in a database. Claims can be positive (i.e., "I am in the database"), negative (i.e., "I am not in the database") or specific (i.e., "I am end user 123 in the database"). 2) When a biometric sample is submitted to a biometric system to verify a claimed identity. 3) The name or index of a claimed reference template or enrollee used by a biometric system for verification.

Claimant - A person submitting a biometric sample for verification or identification while claiming a legitimate or false identity.

Client application - A computer program running on a computer in communication with a card interface device.

Closed-set identification - 1) When an unidentified end-user is known to be enrolled in the biometric system. Opposite of "open-set identification." 2) A biometric task where an unidentified individual is known to be in the database and the system attempts to determine his/her identity. Performance is measured by the frequency with which the individual appears in the system's top rank (or top 5, 10, etc.) *See also identification and open-set identification.*

Coalition - An *ad hoc* arrangement between two or more nations for common action.

Collectability - Refers to how easy or difficult it is to acquire a biometric for measurement.

Common criteria - Standard that provides a comprehensive, rigorous method for specifying function and assurance requirements for products and systems.

Common criteria certification - An international standard for IT security developed jointly by standards bodies in the U.S., Europe, and Canada. Product certification involves a rigorous evaluation process, including both evaluation of product documentation and testing of the product's security related functions.

Common user provisioning (also called one-step provisioning) - Having a single point of employee registration and dismissal (usually in a human resources system) with automatic assignment and revocation of both physical and information security privileges.

Comparison - Process of comparing a biometric reference with a previously stored reference or references in order to make an identification or verification decision; the process of comparing biometric data with a previously stored reference template or templates (biometric data). *See also match.*

Comparison - The process of comparing a biometric sample with a previously stored reference template or templates. *See also one-to-many and one-to-one.*

Compressed template - A template that has been created from a biometric sample by using compression.

Compression - Reducing to a more manageable size the number of bits necessary to represent the relevant information.

Computer network - The constituent element of an enclave responsible for connecting computing environments by providing short-haul data transport capabilities such as local or campus area networks, or long-haul data transport capabilities such as operational, metropolitan, or wide area and backbone networks.

Confidence interval - Refers to the inherent uncertainty in test results owing to small sample size.

Contact/Contactless - In regard to chip cards: whether the card is read by direct contact with a reader or has a transmitter/receiver system which allows it to be read using radio frequency technology (up to a certain distance).

Continuous tone image - An image whose components have more than one bit per sample.

Cookie - A small piece of information (token) sent by a Web server and stored on a user's system (hard drive) so it can later be read back from that system. Using cookies is a convenient technique for having the browser remember specific information. Cookies may be categorized as "session" or "persistent" cookies. "Session" cookies are temporary cookies used to maintain context or "state" between otherwise stateless Web transactions (e.g., to maintain a "shopping basket" of goods selected during a single logical session at a site) and that must be deleted at the end of the web session in which they are created.

"Persistent" cookies remain over time and can be used for a variety of purposes, including to track a user's access over time and across Web sites, or to establish user preferences.

Cooperative user - An individual that willingly provides his/her biometric to the biometric system for capture, example: A worker submits his/her biometric to clock in and out of work. *See also indifferent user, non-cooperative user, and uncooperative user.*

Cooperative versus Non-cooperative - The first partition is "cooperative/non-cooperative". This refers to the behavior of the "wolf", (bad guy or deceptive user). In applications verifying the positive claim of identity, such as access control, the deceptive user is cooperating with the system in the attempt to be recognized as someone s/he is not. This we call a "cooperative" application. In applications verifying a negative claim to identity, the bad guy is attempting to deceptively not cooperate with the system in an attempt not to be identified. This we call a "non-cooperative" application. Users in cooperative applications may be asked to identify themselves in some way, perhaps with a card or a PIN, thereby limiting the database search of stored templates to that of a single claimed identity. Users in non-cooperative applications cannot be relied on to identify themselves correctly, thereby requiring the search of a large portion of the database. Cooperative, but so-called "PINless," verification applications also require search of the entire database.

Core point - The "center(s)" of a fingerprint. In a whorl pattern, the core point is found in the middle of the spiral/

circles. In a loop pattern, the core point is found in the top region of the innermost loop. More technically, a core point is defined as the topmost point on the innermost upwardly curving friction ridgeline. A fingerprint may have multiple cores or no cores. *See also arch, delta point, friction ridge, loop, and whorl.*

Covert - An instance in which biometric samples are being collected at a location that is not known to bystanders. An example of a covert environment might involve an airport checkpoint where face images of passengers are captured and compared to a watch list without their knowledge. *See also non-cooperative user, and overt.*

Credential - Evidence (usually in printed form) concerning one's right to credit, confidence, authority or privileges. Security systems have two categories of credentials used to verify identity and perform authentication of privileges: physical visual credentials (such as a photo ID badge) and electronic credentials (information stored on a security card or in a computer database). Electronic credentials are also called logical credentials. In the context of FIPS 201, the PIV card is a physical credential-a smart card-with specific information printed on it and specific information encoded in the smart card memory. The data encoded on the PIV card is a logical credential.

Credentials - Documents, tokens, information or data, which establish or confirm an individual person's rights and privileges. The production of adequate credentials need not involve disclosure of Identity. Proof of class membership may be necessary and sufficient.

Cross-credentialing - An arrangement between organizations whereby each organization accepts credentials issued by the other. This requires collaboration with regard to many issues including security, privacy, trust, operating rules, policies, and technical standards. The intent of FIPS 201 is to enable cross-credentialing for Federal agencies and their contractors.

Cryptography - Cryptography is the study and practice of protecting information by data encoding and transformation techniques. It includes means of hiding information (such as encryption) and means of proving that information is authentic and has not been altered from its original form (such as digital signatures).

Cryptographic key - A key is a piece of information that controls the operation of a cryptography algorithm. In encryption, a key specifies the particular transformation to be performed on the data being encrypted or decrypted. The key is used to "lock" the data by encrypting it and "unlock" it by decrypting it. Keys are also used in other cryptographic algorithms, such as digital signatures and other schemes for authentication of information.

Cumulative binomial probability distribution - Predicts the probability P of an event occurring k or more times in n trials, where the probability of occurrence of each event is p. In evaluating biometric systems, it is often of interest to determine the probability p, given P, for a comparatively large number of events and trials. Values of P of interest are also often very small or very nearly 1. It happens that this is a difficult computational task if approached in a brute force manner, i.e., performing large numbers of sums for various trial *p*.

Cyber attack - *See also CyI, CyM, CyA, and CyR.*

D

D - Similarity or distance measure

D' - D-Prime - A statistical measure of how well a system can discriminate between a signal and a non-signal.

DA - Department of the Army

DA (also) - Detainee Affairs

DA&M - Director of Administration and Management - Under the direction of the Deputy Secretary of Defense, the DA&M is the principal staff assistant and advisor to the Secretary and Deputy Secretary of Defense on DoD-wide organizational and administrative management matters. These include developing and maintaining organizational charters and overseeing assigned programs, such as the DoD Committee Management Program, DoD Management Headquarters Program, DoD Historical Program, DoD Freedom of Information Act (FOIA) Program, and the OSD Internal Management Control Program. In addition, the DA&M serves as the DoD focal point for DoD Quality Management matters, analyzes and controls manpower requirements for the OSD and other assigned activities, and acts for the Secretary of Defense before the Congressional Joint Committee on Printing on all matters relating to printing, binding, and publications requirements (Title 44, U.S. Code Chapter 11).

DAA - Designated Accrediting Agency - Ultimately responsible for formally assuming responsibility for managing and accepting project risks.

DAA (also) - Designated Approving/Approval Authority

DAB - Defense Acquisition Board

DAC - Defense Acquisition Challenge - Program was established by Congress (Chapter 139 of Title 10 USC, 2359b) to increase the introduction of innovative and cost-saving technologies and products into existing Department of Defense (DoD) acquisition programs. The DAC Program is managed through the Office of Secretary of Defense (Acquisition Technology & Logistics), Deputy Undersecretary of Defense (Advanced Systems & Concepts),

DAC - Discretionary Access Control

DAC (also) - Defense Acquisition Challenge Program

DAE - Defense Acquisition Executive

DAO - Defense Attaché Office

DARPA - Defense Advanced Research Projects Agency - DARPA is a Defense Agency with a unique role within DoD. DARPA is not tied to a specific operational mission. DARPA supplies technological options for the entire Department and is designed to be the "technological engine" for transforming DoD. Near-term needs and requirements generally drive the Army, Navy, Marine Corps, and Air Force to focus on those needs at the expense of major change. Consequently, a large organization like DoD needs a place like DARPA whose only charter is radical innovation. DARPA looks beyond today's known needs and requirements. DARPA's approach is to imagine what capabilities a military commander might want in the future and accelerate those capabilities into being through technology demonstrations. These not only provide options to the commander, but also change minds about what is technologically possible today.

DBEKS - DoD Biometric Expert Knowledgebase System - The DBEKS is a secure website that contains a set of tools designed to enable the coordination of DoD Biometrics with other U.S. Government Agencies regarding biometric activities in the U.S. Government.

DBIDS - Defense Biometric Identification System - A fully configurable security and identification system that enhances safety in an ever-changing world stage of terrorism and conflict. This security is accomplished through accurate identification and access, personal property registration, and fully configurable workstations with a centralized biometric information database. DBIDS was initially developed as a force protection initiative system for the United States Armed Forces, Korea. The system is currently in use at every military installation in Korea. After September 11th, the system was fully implemented to help control access at these installations, where it has been used without interruption.

DCAA - Defense Contract Audit Agency

DCAS - Defense Contract Administration Service

DCCIS - Defense Cross-Credential Identification System - Developed to address specific physical access control needs shared by the DoD and its industry partners. The DCCIS application provides web access to different DCCIS member organization databases, making it possible to authenticate visitors carrying authorized ID cards from fellow DCCIS member organizations.

DCFL - Defense Computer Forensics Laboratory

DCGS-A - Distributed Common Ground System–Army

DCI - Director of Central Intelligence

DCID - Director of Central Intelligence Directive

DCII - Defense Clearance and Investigations Index

DCIN - Defense Continuity Integrated Network

DCIPS - Defense Civilian Intelligence Personnel System

DCIS - Defense Criminal Investigation Service

DCL - Declassify

DCMA - Defense Contract Management Agency

DCMAI - Defense Contract Management Agency International

DCMC - Defense Contract Management Command

DCO - Defense Cooperation Office

DCR - Developed Character Reference

DCS - Defense Courier Service

DCS (also) - Directory of Character Sets

DCSINT - (Army) Deputy Chief of Staff for Intelligence

DD Form 1540 - Registration for Scientific and Technical Information Services

DD Form 1879 - Request for Personnel Security Investigation

DD Form 2024 - DOD Security Classification Guide Data Elements

DD Form 2501 - Courier Authorization Form

DD Form 254 - Contract Security Classification Specification

DD Form 441 - Security Agreement

DDA - The Defence Diversification Agency - Set up by the UK Ministry of Defence in 1999 to promote the spill-over of technology between the civil and defense sectors, is today a leading group in demand-led technology transfer. The DDA extends defense technology to enhance the capability of the UK's armed forces. The DDA has a particular responsibility for assisting small and medium-sized enterprises. The Government considers that significant competitive advantage can be gained for the UK economy from within this responsive, fast moving sector.

DDL - Delegation Disclosure Letter

DDR&E - Deputy Director Research and Engineering

DEA - Data Exchange Agreement

DEA (also) - Drug Enforcement Administration

DEC MAT - Decision Matrixes

DeCA - Defense Commissary Agency

DECL - Declassify

DEPSECDEF - Deputy Secretary of Defense

DERV - Derived From

DES - Data Encryption Standard

DFAR - Defense Federal Acquisition Regulation

DFAS - Defense Finance and Accounting Service

DHS - The United States Department of Homeland Security - Commonly known as Homeland Security, DHS is a Cabinet department of the Federal Government of the United States with the responsibility of protecting the territory of the United States from terrorist attacks and responding to natural disasters. Whereas the Department of Defense is charged with military actions abroad, the Department of Homeland Security works in the civilian sphere to protect the United States within, at, and outside its borders. Its goal is to prepare for, prevent, and respond to domestic emergencies, particularly terrorism. With approximately 184,000 employees, DHS is the third largest cabinet department in the U.S. federal government after the Department of Defense and Department of Veterans Affairs. Homeland security policy is coordinated at the White House by the Homeland Security Council, with Frances Townsend as the Homeland Security Advisor. Other agencies with significant homeland security responsibilities include the Department of Health and Human Services, the Department of Justice, and the Department of Energy.

DHRA - Defense Human Resources Activity

DIA - The Defense Intelligence Agency - DIA is a major producer and manager of military intelligence for the United States Department of Defense. The DIA, designated in 1986 as a Defense Department combat support and intelligence agency, was established in 1961. It was preceded by the Counter Intelligence Corps. Approximately 8000 men and women work for DIA worldwide (about 30% are military personnel and about 70% are civilians). The exact numbers and specific budget information are not publicly released due to security considerations. DIA has major operational activities at the Pentagon, the Defense Intelligence Analysis Center (DIAC), Bolling Air Force Base in Washington, D.C., the Armed Forces Medical Intelligence Center (AFMIC) in Fort Detrick, Maryland, and the Missile and Space Intelligence

Center (MSIC) in Huntsville, Alabama. The DIA is a member of the United States Intelligence Community, reporting to the Director of National Intelligence. The activities of DIA are often compared to Russia's GRU, the UK's Defence Intelligence Staff, and Israel's Aman (IDF).

DIAM - Defense Intelligence Agency Manual

DIAP - Defense-wide Information Assurance Program

DIRNSA - Director, National Security Agency

DIS - Defense Investigative Service (former acronym). *See also DSS.*

DIS FL 381-R - Letter of Notification of Facility Security Clearance

DISA - Defense Information Systems Agency

DISC4 - Director of Information Systems, Command, Control, Communications and Computers

DISCO - Defense Industrial Security Clearance Office

DISCO Form 560 - Letter of Consent

DISCO Form 562 - Personnel Security Clearance Change Notification

DISR - DoD IT Standards Registry

DLSA - Defense Legal Services Agency

DMDC - Defense Manpower Data Center - Archives, and maintains accurate, readily available manpower and personnel data, as well as financial databases for the Department of Defense.

DMRO - Defense Reutilization Management Office

DNA - Deoxyribonucleic Acid - A nucleic acid that contains the genetic instructions for the biological development of a cellular form of life or a virus, all known cellular life and some viruses have DNA. DNA is a long polymer of nucleotides (a polynucleotide) that encodes the sequence of amino acid residues in proteins, using the genetic code.

DNA (also) - Defense Nuclear Agency

DNG - Downgrade

DNI - Director of Naval Intelligence

DNI (also) - The Director of National Intelligence - DNI is the United States government official subject to the authority, direction and control of the President of the United States who is responsible under the Intelligence Reform and Terrorism Prevention Act of 2004 for: 1) Serving as the principal adviser to the President of the United States, the National Security Council, and the Homeland Security Council for intelligence matters related to the national security; and 2) Serving as the head of the sixteen member United States Intelligence Community; and overseeing and directing the National Intelligence Program of the United States.

DNVC - Defense National Visitors Center - DNVC was created in response to a Department of Defense (DoD) mandate that calls for the fortification of methods that DoD organizations use to authenticate DoD identification (ID) card-carrying visitors. It is a web-based system that allows DoD organizations to authenticate credentials and credential holders using photograph, text, and fingerprint data stored in centralized databases. To maintain the privacy and integrity of data transmitted, the DNVC applies industry-standard encryption techniques.

DNVS - Defense National Visitors System

DoC - Department of Commerce

DoD/DOD - The United States Department of Defense - DoD is the federal department charged with coordinating and supervising all agencies and functions of the government relating directly to national security and the military. The DOD is the major tenant of the Pentagon, and it is divided into three major subsections—the U.S. Army, the U.S. Navy, and the U.S. Air Force. Among the many DOD agencies are the Ballistic Missile Defense Organization (see Strategic Defense Initiative), the Defense Advanced Research Projects Agency (DARPA), the Defense Intelligence Agency (DIA), the National Geospatial-Intelligence Agency (NGA), and the National Security Agency (NSA). The department also operates several joint service schools, including the National War College. The United States Coast Guard is not part of the DOD, although it is a branch of military and one of the uniformed services. It is normally under the authority of the United States Department of Homeland Security. However during times of war the Coast Guard can be placed under the authority of the DOD via the Department of the Navy.

DoDAF - Department of Defense Architecture Framework

DODD - Department of Defense Directive

DoDD - Department of Defense Directive

DoDEA - DoD Education Activity

DoDI - Department of Defense Instruction

DoDIG - Department of Defense Inspector General

DoDSI - Department of Defense Security Institute (disestablished)

DOE - Department of Energy

DOHA - Defense Office of Hearings and Appeals (formerly Directorate, Industrial Security Clearance Review (DISCR))

DOI - Digital Object Identifier

DOJ/DoJ - The United States Department of Justice - DOJ is a Cabinet department in the United States government designed to enforce the law and defend the interests of the United States according to the law and to ensure fair and impartial administration of justice for all Americans (see 28 U.S.C. § 501). The DOJ is administered by the United States Attorney General (see 28 U.S.C. § 503), one of the original members of the cabinet.

DOM - Domain Name

DON CAF - Department of Navy Central Adjudication Facility

DOS - Department of State

DOT&E - Department of Test and Evaluation

DOTMLPF - Doctrine, Organization, Training, Materiel, Leadership and Education, Personnel and Facilities

DPD - Delegated Path Discovery

DPI - Dots Per Inch - The number of pixels used to define an image. A measure of image quality is resolution in dots per inch or DPI.

DPMO - Defense Prisoner of War/Missing Personnel Office

DPRB - Defense Planning and Resources Board

DPRO - Defense Plant Representative Office

DPS - Defense Planning Scenarios

DPV - Delegated Path Validation

DRB - Defense Resources Board

DRM - Digital Rights Management

DRS - Detainee Reporting System

DSA - Digital Signature Algorithm - Presented in 1991 by the NIST and patented in 1993. A publicly available one-way algorithm used to generate or verify digital signatures of a text to be signed (not to encrypt/decrypt information). As input, DSA needs: 1) The message digest of the message to be signed; 2) The signer's private key; 3) A random number. Its output is a pair of numbers (often referred to as *r* and *s*), which together, make up the digital signature. To verify a digital signature, DSA needs as input: 1) The message digest of the text to be verified; 2) The signer's public key; 3) The value s from the signature. DSA then makes a computation, the output of which is called *v*, for example. If *v* = *r*, then the signature verifies.

DSB - Defense Science Board - The organization was established in 1956 in response to recommendations of the Hoover Commission. The Board met for the first time on September 20, 1956. Its initial assignment concerned the program and administration of basic research, component research, and the advancement of technology in areas of interest to the Department of Defense. Currently, the Board's authorized strength is thirty-two members and seven ex officio members (the chairmen of the Army, Navy, Air Force, Policy, Defense Business Board and Defense Intelligence Agency advisory committees). The members are appointed for terms ranging from one to four years and are selected on the basis of their preeminence in the fields of science, technology and its application to military operations, research, engineering, and manufacturing and acquisition process.

DSCA - Defense Security Cooperation Agency

DSMC - Defense Systems Management College

DSN - Defense Switched Network

DSS - Defense Security Service

DSS - Defense Security Service - DSS is an agency of the Department of Defense (DoD) located in Alexandria, Virginia with field offices throughout the United States. The Under Secretary of Defense for Intelligence provides authority, direction and control over DSS. DSS provides the military services, Defense Agencies, 23 federal agencies and approximately 12,000 cleared contractor facilities with security support services. DSS administers and implements the defense portion of the National Industrial Security Program (NISP) pursuant to Executive Order 12829. Approximately 412 Industrial Security personnel provide oversight and assistance to cleared contractor facilities and assist management and Facility Security Officers in ensuring the protection of U.S. and foreign classified information. DSS also facilitates classified shipments between the United States and 60 foreign countries and implements foreign ownership, control and influence countermeasures. The DSS Counterintelligence Office increases CI awareness throughout cleared industry and provides suspicious activity reports from industry to 19 other government agencies. The DSS Academy is located in Linthicum, MD, and provides security education and training to DoD security professionals through formal classroom and distributed learning methodologies (i.e., computer-based, web-based and tele-training). The Personnel Security Clearance Office (PSC) consists of the Defense Industrial

Security Clearance Office (DISCO), the Clearance Liaison Office (CLO) and the Polygraph Office. DISCO, located in Columbus, Ohio, processes requests for industrial personnel security investigations and provides eligibility or clearance determinations for cleared industry personnel under the NISP. The CLO is the DoD focal point to the Office of Personnel Management (OPM) on personnel security investigations. The Polygraph Office conducts polygraphs, and provides awareness training requested by central adjudication facilities. DSS provides information technology services in support of DoD and partner agency industrial and personnel security missions.

DSS (also) - Digital Signature Standard Developed by U.S. Federal Information Processing Standard (FIPS). Adopted the DSA in the early 1990s.

DSV - Dynamic Signature Verification - Synonym for signature verification.

DTIC - Defense Technical Information Center

DTIRP - Defense Treaty Inspection Readiness Program

DTRA - Defense Threat Reduction Agency

DTRMC - DoD Test Resource Management Center

DTSA - Defense Technology Security Administration

DUSA - Deputy Under Secretary of the Army

DUSD - Deputy Undersecretary of Defense

DUSD AS&C - Department of the Under Secretary of Defense for Advanced Systems and Concepts

DUSD(TWP) - Deputy Under Secretary of Defense (Tactical Warfare Programs)

DUST - Dual Use Science & Technology - Program designed is to partner with industry to jointly fund the development of dual use technologies needed to maintain our technological superiority on the battlefield and for industry to remain competitive in the marketplace.

Data object - An item of information seen at the card command interface for which are specified a name, a description of logical content, a format, and a coding.

Data vaulting - The process of sending data off site, where it can be protected from hardware failures, theft, and other threats. Several companies now offer web backup services that compress, encrypt, and periodically transmit a customer's data to a remote vault. In most cases, the vaults have auxiliary power supplies, powerful computers, and manned security. Also referred to as a remote backup service (RBS).

Database - A collection of one or more computer files. Any storage of biometric templates and related end user information. For biometric systems, these files could consist of biometric sensor readings, templates, match results, related end user information, etc. *See also gallery.*

Datamining - Technologies generally involve the combination of large volumes of data of various types from many different sources. The potential to connect highly diverse information outside the context of the original collection and to predict characteristics of individuals raises privacy concerns related to data quality and notice. Privacy compliance requirements apply to all uses of PII within a datamining system, which means that the each use of each field of PII should be articulated to facilitate the appropriate level of analysis required to ensure privacy compliance.

Deception - Those measures designed to mislead a foreign power or enemy by manipulation, distortion, or falsification of evidence to induce him to react in a manner prejudicial to his interests or objectives. Deception can be military, political, economic, or scientific and technological (S&T) at national or strategic or lesser levels.

Deception means - Methods, resources, and techniques that can be used to convey information to the deception target. There are three categories of deception means: 1) physical – activities and resources used to convey or deny selected information to a foreign power (e.g., military operations, including exercises, reconnaissance, training activities, and movement of forces. The use of decoy equipment and devices; tactics; bases, logistic actions, stockpiles, and repair activity; and test and evaluation activities); 2) technical – military material resources and their associated operating techniques used to convey or deny selected information to a foreign power through the deliberate radiation, re-radiation, alteration, absorption, or reflection of energy; the emission or suppression of chemical or biological odors; and the emission or suppression of nuclear particles; and 3) administrative

– resources, methods, and techniques to convey or deny oral, pictorial, documentary, or other physical evidence to a foreign power.

Deception methods - The procedures, techniques, or processes for conveying deception information to a target and/or causing the formulation of a particular perception of reality. Deception methods have been subdivided into: 1) Fabrication – The creation of totally false or fictitious evidence or information; 2) Manipulation – The mixing of factual and fictitious or exaggerated evidence; 3) Conditioning – The repeated presentation or transmission of signals, information, or activities to a foreign or target intelligence service; and 4) Diversion – The act of drawing an adversary's attention away from an area, event, or activity of importance to the perpetrator toward on of less importance. Desensitizing its analytical elements to the point where they are no longer viewed with concern or alarm.

Decision - 1) The result of the comparison between the score and the threshold. 2) The decisions a biometric system can make include match, non-match, and inconclusive, although varying degrees of strong matches and non-matches are possible. Either/or multimodality describes systems that offer multiple biometric technologies but only require verification through a single technology. 3) The resultant action taken (either automated or manual) based on a comparison of a similarity score (or similar measure) and the system's threshold. 4) A determination of probable validity of a user's claim to identity/non-identity in the system. *See also comparison, similarity score, and threshold.*

Decision policy - A criteria and process which the evaluator use to determine that two iris images match or do not match (e.g., decision based on threshold and retrying capture.)

Decryption - The changing of encrypted information back into readable form using a decryption key digital certificate.

Defense Production Act Title III - The Title III Program is a DoD-wide initiative that establishes, maintains or expands a production capability offered for national defense. Management responsibilities include program oversight and guidance, strategic planning and legislative proposals, approval of new projects, and liaison with other Federal agencies and Congress.

Degrees of freedom - A statistical measure of how unique biometric data is. Technically, it is the number of statistically independent features (parameters) contained in biometric data.

Delta point - Part of a fingerprint pattern that looks similar to the Greek letter delta. Technically, it is the point on a friction ridge at or nearest to the point of divergence of two type lines, and located at or directly in front of the point of divergence. *See also core point and friction ridge.*

Design data - Data used to optimize the design and parameters of the verification algorithm, verification device, authentication system, identification algorithm, identification device, and individual identification system. Examples include 1) A face images or features used in the design of a face dictionary; and 2) A face images used to determine the parameters of the facial-part extraction.

Detection and identification rate - The rate at which individuals, who are in a database, are properly identified in an open-set identification (watch list) application. *See also open-set identification, and watch list.*

Detection error trade-off (DET) curve - A graphical plot of measured error rates, DET curves typically plot matching error rates (false non-match rate vs. false match rate) or decision error rates (false reject rate vs. false accept rate). *See also receiver operating characteristics.*

Device - Biometric sensor hardware used to capture raw biometric samples from a subject.

Difference score - A value returned by a biometric algorithm that indicates the degree of difference between a biometric sample and a reference. *See also hamming distance and similarity score.*

Digital certificate - 1) In the PKI environment, the data, equivalent to an identity card, issued to a user by a CA (certificate authority), which is used during business transactions to prove his/her identity. Sometimes called a digital ID. 2) The electronic counterpart to a driver license, passport or membership card. It is specially formatted block of data that serves as a form of personal identification that can be verified electronically. A digital certificate is what binds a public key to an identity (a person or system) and is a means of establishing trust in electronic communications. The certificate is issued by a trusted authority (called the certificate authority). This authority stores the digital certificates it publishes in a computer database or network directory, which it makes available online (in a local area network or on the Internet) so that software applications can verify digital signatures as needed. Certificate verification is performed automatically by the software of systems that use digital certificates for information protection (such as e-mail systems).

Digital certificate subscriber - The person to whom a digital certificate is issued, usually simply referred to as the "subscriber" in discussions about digital certificates.

Digital communications - The use of electronic digital signals (ones and zeros) to send information between electronic devices or systems using wired, wireless (radio) or fiber-optic means of transmission.

Digital signature - 1) The number derived by performing cryptographic operations on the text to be signed. This operation, or hash function (also called hash algorithm), is performed on the binary code of the text. The result is known as the message digest, and always has a fixed length. A signature algorithm is applied to the message digest, resulting in the digital signature. 2) Transformation of a message using an asymmetric cryptosystem so that a person who has the initial message and the signer's public key can accurately determine: a) whether the transformation was created using the private key that corresponds to the signer's public key; and b) whether the initial message has been altered since the transformation was made.

Dimension - Number of pixels in either *x*- or *y*-direction

Discrete type - An implementation type in a verification device and an authentication system. In case of a verification device, the iris-imaging function and matching function are installed in separate units. In case of an authentication system, individual functions are installed in separate units.

Discriminant training - A means of refining the extraction algorithm so that biometric data from different individuals are as distinct as possible.

Disruptor - A person or persons who disrupts or attempts to disrupt the correct operation of a biometric system for human ID or ID claim verification. Successful disruption leads to the system providing false decisions or scores to the application or failing to operate.

DoD information system - Set of information resources organized for the collection, storage, processing, maintenance, use, sharing, dissemination, disposition, display, or transmission of information. Includes automated information system applications, enclaves, outsourced IT-based processes, and platform IT interconnections.

Double-dipping - Occurs when people claim benefits under multiple identities. Biometric systems can screen applicants to determine if they are already enrolled under one name and prohibit enrollment in multiple accounts

Duplicate search - Computer generated search performed to detect any duplicated stored in a permanent database.

Dynamic biometric verification method - A biometric verification method, which requires a dynamic action from the person to be authenticated (i.e., a user response to a challenge).

E

E.O. - Executive Order - U.S. Presidents have issued executive orders since 1789. There is no Constitutional provision or statute that explicitly permits this, aside from the vague grant of "executive power" given in Article II, Section 1 of the Constitution and the statement "take care that the laws be faithfully executed" in Article II, Section 3. Most executive orders are orders issued by the President to U.S. executive officers to help direct their operation, the result of failing to comply being removal from office.

EAA - Export Administration Act of 1979, 50 U.S.C. App. 2401 *et. seq.*

EAL - Evaluation Assurance Level

EAR - Export Administration Regulations, 15 CFR 768-799

EBTS - Electronic Biometric Transmission Specification

ECB - Electronic Code Book

ECC - Elliptic Curve Cryptography

ECCM - Electronic Counter-Countermeasures

ECDSA - Elliptic Curve Digital Signature Algorithm

ECM - Electronic Countermeasures

EDI-PI - Electronic Data Interchange Personal Identifier - 1) The EDI-PI is a unique system identifier that is used for machine-to-machine transactions by the Department of Defense. In the Defense Enrollment Eligibility Reporting System (DEERS), the central repository for DoD person data, the EDI-PI is used as the primary identifier for all individuals. It is not a number that is known to the individuals, and it is never intended that the EDI-PI be used outside of machine-to-machine transactions. 2) The EDI-PI is the personal unique identifier used as part of the Cardholder Unique Identifier, which is part of the Homeland Security Presidential Directive-12 solution for the Department of Defense. As such, it may be used as an identifier when the Common Access Card is used to electronically authenticate an individual. A greater shift to electronic authentication would reduce the use of the SSN and provide greater security for transactions.

EEFI - Essential Elements of Friendly Information

EEI - Essential Elements of Information

EEPROM - Electrically Erasable Programmable Read-Only Memory

EFTS - Electronic Fingerprint Transmission Specification - A document that specifies requirements to which agencies must adhere to communicate electronically with the

Federal Bureau of Investigation's Integrated Automated Fingerprint Identification System (IAFIS). This specification facilitates information sharing and eliminates the delays associated with fingerprint cards. *See also IAFIS.*

ELECTRO-OPINT - Electrical Optical Intelligence

ELINT - Electronic Intelligence

ELSEC - Electronic Security

EMIO - Enhanced Maritime Interdiction Operations - System to collect and match biometric data obtained from individuals encountered during Visit, Board, and Search and Seizure (VBSS) operations. Rapid intelligence exploitation is key.

EMSEC - Emissions Security

ENAC - Entrance National Agency Check

ENTNAC - Entrance National Agency Check - The primary reason for the ENTNAC is to determine the suitability of an individual for entry into the military.

EOC - Emergency Operations Center

EPL - Evaluated Products List

EPSQ - Electronic Personnel Security Questionnaire

EPW - Enemy Prisoner of War

ESI - Extremely Sensitive Information

ETIS - Enhanced Terrorist Identification Service

ETSI - European Telecommunications Standards Institute

EQIP - Electronic Questionnaires for Investigations Processing

EUCOM - U.S. European Command

EXCOM - Executive Committees

Ear shape - A lesser-known physical biometric that is characterized by the shape of the outer ear, lobes, and bone structure.

Eavesdropping - Surreptitiously obtaining data from an unknowing end user who is performing a legitimate function. An example involves having a hidden sensor co-located with the legitimate sensor. *See also skimming.*

Eigenface - A method of representing a human face as a linear deviation from a mean or average face.

Eigenhead - The three dimensional version of Eigenface that also analyses the shape of the head.

Electronic credential - Information stored on a security card or in a computer database as evidence of privileges or authority. *See also credential.*

Electronic form - An officially prescribed set of data residing in an electronic medium that is used to produce as near to a mirror-like image as the creation software will allow of the officially prescribed form. An electronic form can also be one in which prescribed fields for collecting data can be integrated, managed, processed, and/or transmitted through an organization's IT system. There are two types of electronic forms: one that is part of an automated transaction, and one whose image and/or data elements reside on a computer.

Encryption - The scrambling of data so that it becomes difficult to unscramble or decipher. Scrambled data is called ciphertext, as opposed to unscrambled data, which is called plaintext. Unscrambling ciphertext is called decryption. Data encryption is done by the use of an algorithm and a key. The key is used by the algorithm to scramble and unscramble the data. The algorithm can be public (for scrutinization and analysis by the cryptographic community), but the key must be kept private. Encryption does not make unauthorized decryption impossible, but merely difficult. Time, and the power (ever increasing) of computers are the factors involved in the feasibility of decryption; Transforming a test into code to conceal its meaning. The process of transforming data to an unintelligible form so that the original data either cannot be obtained (one-way encryption) or cannot be obtained without using the inverse decryption process

End user - 1) The individual who will interact with the system to enroll, to verify, or to identify. 2) A person who interacts with a biometric system to enroll or have his/her identity checked. *See also cooperative user, indifferent user, non-cooperative user, and uncooperative user.*

End user adaptation - The process of adjustment whereby a participant in a test becomes familiar with what is required and alters their responses accordingly.

Enrollee - A person who has a biometric reference template stored in a biometric package.

Enrollment - 1) The process of collecting a biometric sample from an end user, converting it into a biometric reference, and storing it in the biometric system's database for later comparison. 2) The process of collecting biometric samples from a user and the subsequent preparation, encryption, and storage of biometric reference templates representing that person's identity. 3) The initial process of collecting biometric data from a user and then storing it in a template for later comparison; the process of collecting a biometric sample(s) from an individual, and the subsequent construction of a reference template(s) representing the individual's identity.

Enrollment data - Data that indicates the characteristics of a biometric to be registered prior to matching or matching decision. Generally, this data is created from multiple features of a fingerprint, for example, and is also referred to as a template.

Enrollment data set - An enrollment data set is a collection of one or more pieces of enrollment data.

Enrollment incompatibility - Indicates that some types of signature data or test volunteers cannot be enrolled. In addition, it indicates that some types of signature data or test volunteers cannot be added.

Enrollment process - 1) Raw data acquisition from sensor. 2) Feature extraction raw data and biometric reference data in the card

Enrollment time - 1) The time period a person must spend to have his/her biometric reference template successfully created. 2) Time required of an enrollee, after initial orientation, to successfully create his/her biometric reference template.

Enrollment unavailability - Indicates that some types of voice data or test volunteers (speakers) cannot be enrolled, such as because enrollment data cannot be generated. In addition, it indicates that some types of voice data or test volunteers (speakers).

Enrollment voice data - Voice data to be used for enrollment.

eSignature - eSignature (or signature recognition) systems authenticate the identity of individuals by measuring their hand-written signatures. Signature recognition measures how the signature is signed, rather than comparing signatures after they have been written. eSignature systems analyze the dynamics of a signature, including acceleration rates, directions, pressure, and stroke order and count.

Ethernet - Ethernet is a local-area network (LAN) protocol developed by Xerox Corporation in cooperation with DEC and Intel in 1976. It is one of the most widely implemented LAN standards.

Equal Error Rate (EER) - A statistic used to show biometric performance, typically when operating in the verification task. The EER is the location on a ROC or DET curve where the false accept rate and false reject rate (or one minus the verification rate {*1 - VR*}) are equal. In general, the lower the equal error rate value, the higher the accuracy of the biometric system. However, that most operational systems are not set to operate at the "equal error rate" so the measure's true usefulness is limited to comparing biometric system performance. The EER is sometimes referred to as the "Crossover Error Rate." *See also Detection Error Trade-off (DET) curve, false accept rate, false reject rate, and Receiver Operating Characteristics (ROC).*

Explicit claim of identity - In applications where there is an explicit claim of identity or non-identity, the submitted sample needs to be matched against just the enrolled template for that identity. The accept/reject decision depends on the result of a one-to-one comparison.

Extraction - 1) The process of converting a captured biometric sample into biometric data so that it can be compared to a reference template. 2) Process by which the biometric sample captured in the previous block is transformed into an electronic representation. During enrollment, this electronic representation is known as the biometric template. During the authentication process, it is known as the live sample. 3) The process of converting a captured biometric sample into biometric data so that it can be compared to a reference. 4) Process by which the biometric sample captured during enrollment is transformed into an electronic representation *See also biometric sample, feature, and template.*

Eye scans - Eye scans can be categorized into two types: iris and retinal. Iris scans digitally process, record, and compare the light and dark patterns in the iris' flecks and rings, something akin to a human bar code. Some claim this technique is more accurate than a fingerprint and can be employed at such a distance that the person being scanned is unaware. Others say these systems can easily be fooled. Researchers testing one system discovered that university students who wore patterned "designer" contacts were wrongly rejected because the contacts were in a different position every time the students' eyes were scanned. Retinal scans, on the other hand, are more intrusive, requiring close-up infrared scanning through the pupil.

F

FAA - Functional Area Analysis

FAR - False Acceptance Rate - 1) The percentage of imposters incorrectly matched to a valid user's biometric; a statistic used to measure biometric performance when operating in the verification task. The percentage of times a system produces a false accept, which occurs when an individual is incorrectly matched to another individual's existing biometric. Example: Frank claims to be John, and the system verifies the claim.

FAR (also) - False Alarm Rate - A statistic used to measure biometric performance when operating in the open-set identification (sometimes referred to as watch list) task. This is the percentage of times an alarm is incorrectly sounded on an individual who is not in the biometric system's database (the system alarms on Frank when Frank isn't in the database), or an alarm is sounded but the wrong person is identified (the system alarms on John when John is in the database, but the system thinks John is Steve). *See also false match rate, and type II error.*

FAR (also) - Federal Acquisition Regulation

FASC-N - Federal Agency Smart Credential Number - The Federal Agency Smart Credential Number (FASC-N) is one of the data items contained within the CHUID, and uniquely identifies a PIV card. The FASC-N replaces the SEIWG-012 definition, which has been in use for over 10 years.

FBCA - Federal Bridge Certification Authority

FBI - Federal Bureau of Investigation - The FBI is the investigative arm of the U.S. Department of Justice. The FBI's investigative authority can be found in Title 28, Section 533 of the U.S. Code. Additionally, there are other statutes, such as the Congressional Assassination, Kidnapping, and Assault Act (Title 18, U.S. Code, Section 351), which give the FBI responsibility to investigate specific crimes.

FBIS - Foreign Broadcast Information Service

FCL - Facility Security Clearance

FCS - Future Combat System - The Army's Future Combat Systems network allows the FCS Family-of-Systems (FoS) to operate as a cohesive system-of-systems where the whole of its capabilities is greater than the sum of its parts. As the key to the Army's transformation, the network, its logistics, and Embedded Training (ET) systems, enable the Future Force to employ revolutionary operational and organizational concepts. The network enables Soldiers to perceive, comprehend, shape, and dominate the future battlefield at unprecedented levels as defined by the FCS Operational Requirements Document (ORD). The FCS network consists of four overarching building blocks: System-of-Systems Common Operating Environment (SOSCOE); Battle Command (BC) software; communications and computers (CC); and intelligence, reconnaissance and surveillance (ISR) systems. The four building blocks synergistically interact enabling the Future Force to see first, understand first, act first and finish decisively. System-of-Systems Common Operating Environment (SOSCOE) Central to FCS network implementation is the SOSCOE, which supports multiple mission-critical applications independently and simultaneously. It is configurable so that any specific instantiation can incorporate only the components that are needed for that instantiation. SOSCOE enables straightforward integration of separate software packages, independent of their location, connectivity mechanism and the technology used to develop them. SOSCOE architecture uses commercial off-the-shelf hardware and a Joint Tactical Architecture-Army compliant operating environment to produce a non-proprietary, standards-based component architecture for real-time, near-real-time, and non-real-time applications. SOSCOE also contains administrative applications that provide capabilities including login service, startup, logoff, erase, memory zeroize, alert/emergency restart and monitoring/control. SOSCOE framework allows for integration of critical interoperability services that translate Army, Joint, and coalition formats to native, internal FCS message formats using a common format translation service. Because all interoperability services use these common translation services, new external formats will have minimal impact on the FCS software baseline. The FCS software is supported by application-specific interoperability services that act as proxy agents for each Joint and Army system. Battle Command (BC) can access these interoperability services through application program interfaces that provide isolation between the domain applications, thereby facilitating ease of software modifications and upgrades.

FCT - Foreign Comparative Testing Program - The mission is to test items and technologies of foreign allies and friends that have a high Technology Readiness Level (TRL) in order to satisfy valid defense requirements more quickly and economically. Within the FCT Program, foreign items are nominated by a sponsoring organization within the Department of Defense for testing in order to determine whether the items satisfy U.S. military requirements or address mission area shortcomings. The OSD Comparative Testing Office funds testing and evaluation; the Services fund all procurements that result from a successful test. The FCT Program's objectives are to improve the U.S. warfighter's capabilities and reduce expenditures through: 1) Rapidly fielding quality military equipment; 2) Eliminating unnecessary duplication of research, development, test, and evaluation; 3) Reducing life cycle or procurement costs; 4) Enhancing standardization and interoperability; 5) Promoting competition by qualifying alternative sources; and 6) Improving the U.S. military industrial base.

FD Form 258 - Applicant Fingerprint Card

FDAU - Fraudulent Document Analysis Unit

FDL - Forensic Document Laboratory

FEC - Federal Election Commission

FEMA - Federal Emergency Management Agency

FERET - FacE REcognition Technology program - A face recognition development and evaluation program sponsored by the U.S. Government from 1993 through 1997.

FFRDC - Federally Funded Research and Development Center

FG - Focus Group

FGI - Foreign Government Information

FGP - Finger Position

FIDS - Facility Intrusion Detection System

FIM - Federated Identity Management - A number generated by applying a mathematical formula (an

algorithm) to a document or sequence of text, used
for verifying that the document has not been changed
since the original hash value was generated. A hash is
significantly shorter that the original text. The hash number
is unique to the original document, thus attaching it to
a document has negligible impact on the overall size of
the document. The algorithm works one-way: it yields
the same hash result every time for the same message,
and it is not possible in practice for a message to be
reconstituted from the hash result. Also, two different
messages cannot produce the same hash results. Thus if
the sender creates a hash for a document and provides it
to the recipient of the document, the recipient (applying the
same algorithm) can create a hash value and verify that
the hash is identical to the sender's hash, which means
that the document has not been altered. Hashes are used
in the creation of digital signatures.

FIPP - Fair Information Practice Principle

FIPS - Federal Information Processing Standards

FIPS 201 - Federal Information Processing Standard
(FIPS) Publication 201, commonly known by the shorter
name FIPS 201, is titled "Personal Identity Verification
(PIV) of Federal Employees and Contractors." It is both
a standard and a specification. FIPS 201 specifies the
architecture and technical requirements for a common
identification standard for Federal employees and
contractors. The overall goal is to achieve appropriate
security assurance for multiple applications by efficiently
verifying the claimed identity of individuals seeking
physical access to Federally controlled government

facilities and electronic access to government information systems. The standard contains two major sections. Part one describes the minimum requirements for a Federal personal identity verification system that meets the control and security objectives of Homeland Security Presidential Directive 12, including personal identity proofing, registration, and issuance. Part two provides detailed specifications that will support technical interoperability among PIV systems of Federal departments and agencies. It describes the card elements, system interfaces, and security controls required to securely store, process, and retrieve identity credentials from the card. The physical card characteristics, storage media, and data elements that make up identity credentials are specified in this standard. The interfaces and card architecture for storing and retrieving identity credentials from a smart card are specified in Special Publication 800-73, Interfaces for Personal Identity Verification. Similarly, the interfaces and data formats of biometric information are specified in Special Publication 800-76, Biometric Data Specification for Personal Identity Verification. This standard does not specify access control policies or requirements for Federal departments and agencies.

FIQM - Finger Image Quality Measurement

FIRS – Fingerprint Identification Records System

FISINT - Foreign Instrumentation Signals Intelligence

FISMA - Federal Information Security Management Act

FISS - Foreign Intelligence Security Service

FIU - Fingerprint Identification Unit - A biometric system capable of capturing, storing, and comparing fingerprint data for the purposes of verifying an individual's identity.

FiXs - Federation for Identity and Cross-Credentialing Systems - FiXs is a coalition of commercial companies, government contractors, and not-for-profit organizations whose mission is to establish and maintain a worldwide, interoperable identity and cross-credentialing network built on security, privacy, trust, standard operating rules, policies, and technical standards. The FiXs network verifies and authenticates the identity of personnel seeking to enter United States military installations and other government-controlled areas, as well as commercial sites tied to the network.

Flash badge/pass - Identification badge that is visually verified upon entry and/or exit from a facility.

FM&C - Financial Management and Comptroller

FMR(ô) - false match rate: the probability that a sample will be mistakenly matched with a non-self template.

FMS - Foreign Military Sales

FNA - Functional Needs Analysis

FNM*i* - The probability that the *i*th sample will be falsely not matched because of binning or matching errors.

FNMR - False Non-Match Rate - Alternative term for False Rejection Rate, used in the context of Negative Claim of Identity.

FOC - Final Operating Capacity

FOCI - Foreign Ownership, Control, or Influence

FOIA - Freedom of Information Act

FOIA request - A written Freedom of Information Act request for DoD records that reasonably describes the record(s) sought, made by any person, including a member of the public (U.S. or foreign citizen/entity), an organization, or a business, but not including a Federal Agency or a fugitive from the law, that either explicitly or implicitly invokes the FOIA, DoD Directive 5400.7, or DoD Component supplementing regulations or instructions. Requesters should also indicate a willingness to pay fees associated with the processing of their request or, in the alternative, why a waiver of fees may be appropriate. Written requests may be received by postal service or other commercial delivery means, by facsimile, or electronically. Requests received by facsimile or electronically must have a postal mailing address included since it may not be practical to provide a substantive response electronically. The request is considered properly received or perfected, when the above conditions have been met and the request arrives at the FOIA office of the Component in possession of the records.

FOUO - For Official Use Only

FP - Force Protection

FPI - Fixed Price Incentive

FPI (also) - Force Protection Initiative

FPIF - Fixed Price Incentive Firm

FPO - Fleet Post Office

FpVTE - Fingerprint Vendor Technology Evaluation - An independently administered technology evaluation of commercial fingerprint matching algorithms.

FRAC - Foreseeable Risk Analysis Center

FRD - Formerly Restricted Data

FRGC - Face Recognition Grand Challenge - A face recognition development program sponsored by the U.S. Government from 2003-2005. *See also FERET and FRVT.*

FRR - False-Rejection Rate - The percentage of incorrectly rejected valid users.

FRUS - Foreign Relations of the United States

FRVT - Face Recognition Vendor Test - A series of large-scale independent technology evaluations of face recognition systems. *See also FRGC and FERET.*

FSA - Functional Solution Analysis

FSO - Facility Security Officer

FSTS - Federal Security Telephone Service

FTA - Failure to Acquire - Failure of a biometric system to capture and/or extract usable information/data from a biometric sample.

FTE - Failure to Enroll - 1) Failure of a biometric system to form a proper enrollment reference for an end user. Common failures include end users who are not properly trained to provide their biometrics, the sensor not capturing information correctly, or captured sensor data of insufficient quality to develop a template. 2) Any irrecoverable failure in the enrollment process.

FTE rate - The probability that a biometric system will have a failure-to-enroll.

FTTTF - Foreign Terrorist Tracking Task Force

FVS - Foreign Visits System

FY - Fiscal Year

FYDP - Future Year Defense Plan

Face monitoring - A biometric application of face recognition technology where the biometric system monitors the attendance of an end user. This may be overt or covert.

Face recognition - A biometric modality that uses an image of the visible physical structure of an individual's face for recognition purposes.

Facial thermogram - A specialized face recognition technique that senses heat in the face caused by the flow of blood under the skin.

Failure to Acquire - FTA - Failure of a biometric system to capture and extract biometric/comparison data.

Failure to Enroll - FTE - Failure of the biometric system to form a proper enrollment template for an end-user, the failure may be due to failure to capture the biometric sample or failure to extract template data (of sufficient quality).

False acceptance - When a biometric system incorrectly identifies an individual or incorrectly authenticates an imposter against a claimed identity. A type II error.

False match - A match decision when the individual has made an impostor claim of identity.

False Match Rate (FMR) - A statistic used to measure biometric performance when, similar to the False Acceptance Rate (FAR).

False non-match - A statistic used to measure biometric performance. Similar to the False Reject Rate (FRR), except the FRR includes the Failure to acquire error rate and the False Non-Match Rate does not. A non-match decision when the individual has made a genuine or defrauder claim of identity. The rate for incorrect negative matches by the matching algorithm for single template comparison attempts. For a biometric system that uses just one attempt to decide acceptance, FNMR is the same as FRR.

False rejection - 1) When a biometric system fails to identify an enrollee or fails to verify the legitimate claimed identity of an enrollee; when a biometric system rejects a genuine claim on identity of an individual. In both a verification type system and an identification type system, a false rejection occurs as a result of a false non-match. 2) A failure to identify or verify a genuine enrollee. 3) Type I error.

False Rejection Rate (FRR) - A statistic used to measure biometric performance when operating in the verification task. The percentage of times the system produces a false reject. A false reject occurs when an individual is not matched to his/her own existing biometric template. Example: John claims to be John, but the system incorrectly denies the claim. *See also false non-match rate, type I error.*

Feature extraction - 1) The process of converting a captured biometric sample into biometric feature data so that it can be compared to a reference template. 2) The automated process of locating and encoding distinctive characteristics from a biometric sample in order to generate a template. *See also extraction.*

Feature parameter - Numerical data that characterize biometric data, generated from biometric data by some kind of processing.

Feature(s) - 1) Distinctive mathematical characteristic(s) derived from a biometric sample. Used to generate a reference. 2) A mathematical representation of the information extracted from the presented sample by the signal processing subsystem that will be used to construct or compare against enrollment templates (e.g., minutiae coordinates). 3) Data inherent in a fingerprint extracted from fingerprint images. 4) Data specific to each iris image, obtained through specific image processing. *See also extraction and template.*

Features type - An input type in a matching-algorithm and a verification device.

Field test/trial - A trial of a biometric application in "real world" as opposed to laboratory conditions.

Filtering - 1) The process of classifying biometric data according to information that is unrelated to the biometric data itself. This may involve filtering by sex, age, hair color or other distinguishing factors, and including this information in an end user's database record. This term is particularly used in conjunction with Automated Fingerprint Identification Systems. A specialized technique used by some AFIS vendors. 2) Process of classifying finger images according to data which is unrelated to the finger image itself.

Finger geometry - A physical biometric that analyses the shape and dimensions of one or more fingers.

Fingerprint - The image left by the minute ridges and valleys found on the hand of every person. In the fingers and thumbs, these ridges form patterns of loops, whorls, and arches.

Fingerprint recognition - A biometric modality that uses the physical structure of an individual's fingerprint for recognition purposes. Important features used in most fingerprint recognition systems are minutiae points that include bifurcations and ridge endings. *See also bifurcation, core point, delta point, and minutia(e) point.*

Fingerprint scanning - Acquisition and recognition of a person's fingerprint characteristics for identifying purposes. This process allows the recognition of a person through quantifiable physiological characteristics that detail the unique identity of an individual.

Fingerprint sensor - Part of a biometric device used to capture a fingerprint image for subsequent processing.

Fixed-text system - *See text dependent system.*

Forensic DNA analysis - The science of using molecular techniques to analyze specific portions of an individual's DNA in order to establish identity. Three different types of DNA analyses can be performed using the nuclear DNA and mitochondrial DNA (mtDNA) present in human cells to generate a complete genetic profile.

Form - A fixed arrangement of captioned spaces designed for entering and extracting prescribed information. Forms may be preprinted paper forms or electronic forms.

Free-text system - *See text independent system.*

Friction ridge - The ridges present on the skin of the fingers and toes, and on the palms and soles of the feet, which make contact with an incident surface under normal touch. On the fingers, the distinctive patterns formed by the friction ridges that make up the fingerprints. *See also minutia(e) point.*

G

G-2 - Staff Intelligence Officer

GAO - General Accounting Office

GBA - Generic Bootstrapping Architecture

GCA - Government Contracting Activity

GDIP - General Defense Intelligence Programs

GFE - Government Furnished Equipment

GFP - Government Funded Property

GIG - Global Information Grid - GIG is defined as the globally interconnected, end-to-end set of information capabilities, associated processes, and personnel for collecting, processing, storing, disseminating, and managing information on demand to warfighters, policymakers, and support personnel. The GIG includes all owned and leased communications and computing systems and services, software (including applications), system data, security services, and other associated services necessary to achieve information superiority for the United States military. It is the physical manifestation of the network-centric warfare doctrine. The GIG was envisioned by the Department of Defense Chief Information Officer on September 22, 1999 and was officially mandated by an overarching directive from the Deputy Secretary of Defense on September 19, 2002. Noteworthy progress has been made since then. Although the lofty objective of the Global Information Grid has not yet been realized, computer-enabled communication between soldiers and commanders in the battlefield have been successful, most notably during the 2003 invasion of Iraq. This ability is considered an early GIG component.

GOTS - Government Off-the-Shelf

GPS - Global Positioning System

GSA - General Security Agreement

GSA - General Services Administration

GSC-IAB - Government Smart Card Interagency Advisory Board

GSC-IS - Government Smart Card Interoperability Specification

GUID - Global Unique Identification Number

GWOT - Global War on Terror

Gait - An individual's manner of walking. This behavioral characteristic is in the research and development stage of automation.

Gallery - The biometric system's database or set of known individuals, for a specific implementation or evaluation experiment. *See also database and probe.*

Genetic penetrance - The degree to which characteristics are passed from generation to generation.

Genuine attempt - A single good faith attempt by a user to match his or her own stored template.

Genuine claim of identity - A user making a truthful positive claim about identity in the system, the user truthfully claims to be himself/herself, leading to a comparison of a sample with a truly matching template.

Geospatial - Geospatial technologies involve the use of geographic information. Since every object and every individual is located somewhere, geospatial technologies can serve as a universal link between all other information, objects, events, and individuals. The ability to associate location with an individual over time along with all other objects and events associated with the same location raises privacy concerns related to tracking and profiling.

Goat - Biometric system end user whose pattern of activity when interfacing with the system varies beyond the specified range allowed by the system and who consequently may be falsely rejected by the system.

H

HAAPI - Human Authentication Application Programming Interface

HA/DR - Humanitarian Assistance/Disaster Relief

HAC - House Appropriations Committee

HASC - House Armed Services Committee

HIIDE - Handheld Interagency Identity Detection Equipment

HNSC - House National Security Committee

HOF - Home Office Facility

HOIS - Hostile Intelligence Services

HPSCI - House Permanent Select Committee on Intelligence

HQDA - Headquarters, Department of the Army - HQDA is the executive part of the Department of the Army at the seat of Government. It is the highest-level headquarters in the Department and exercises directive and supervisory control over it. HQDA is composed of the Office of the Secretary of the Army; Office of the Chief of Staff, Army; the Army Staff; and specifically designated staff support agencies. It is not restricted to agencies and personnel located in the Washington DC metropolitan area, but includes dispersed agencies and personnel performing "national headquarters" functions, as distinguished from "field" or "local" functions. Within Army regulations, those support and reporting responsibilities set aside for MACOMs generally apply to HQDA, unless otherwise specified.

HSDI - Homeland Security Defense Initiative

HSPD - Homeland Security Presidential Directive

HSPD-12 - Homeland Security Presidential Directive 12 - In 2004, a Homeland Security Presidential Directive

(HSPD) was issued entitled HSPD-12 "Policy for a Common Identification Standard for Federal Employees and Contractors." HSPD-12 establishes the requirement for a mandatory Government-wide standard for secure and reliable forms of identification issued by the Federal Government in order to enhance security, increase Government efficiency, reduce identify fraud, and protect personal privacy, and directed the promulgation of a new Federal standard for secure and reliable identification. This impacts Federal Department and Agency employee and contractors who require long-term access to Federally controlled facilities and information systems. This includes the Department of Defense, Department of State, Armed Forces, Foreign Service, U.S. Postal Service, and all other executive branch components. Government Corporations are encouraged to comply but are not required.

HUMINT - HUMan INTelligence - 1) A category of intelligence gathering disciplines that encompasses all gathering of intelligence by means of interpersonal contact. 2) A category of intelligence derived from information collected and provided by human sources. Most HUMINT activity does not involve clandestine or covert activities. The manner in which HUMINT operations are conducted is dictated by both official protocol and the nature of the source, who may be witting, unwitting, neutral, friendly or hostile. Examples of HUMINT sources include, but are not limited to, the following: a) friendly forces (Military police, patrols, etc.); b) Prisoners of War (POW) or detainees; c) refugees; d) civilians; e) non-governmental organizations (NGOs); f) media personnel/organizations; g) covert/ clandestine agents; and h) walk-ins (someone who approaches a friendly agency and volunteers to provide information on his/her own freewill).

Habituated versus Non-habituated - Applies to the intended users of an application. Users presenting a biometric trait on a daily basis can be considered habituated after a short period of time. Users who have not presented the trait recently can be considered "non-habituated". A more precise definition will be possible after we have better information relating system performance to frequency of use for a wide population over a wide field of devices. If all the intended users are "habituated," the application is considered a habituated application. If all the intended users are "nonhabituated," the application is considered non-habituated. In general, all applications will be non-habituated during the first week of operation, and can have a mixture of habituated and non-habituated users at any time thereafter. Access control to a secure work area is generally habituated. Access control to a sporting event is generally nonhabituated.

Hamming distance - The number of non-corresponding digits in a string of binary digits. Used to measure dissimilarity. Hamming distances are used in many Daugman iris recognition algorithms. *See also difference score and similarity score.*

Hand geometry recognition - A biometric modality that uses the physical structure of an individual's hand for recognition purposes.

Hash function - A function that maps a variable-length data block or message into a fixed-length value called a message digest or hash code. The function is designed so that, when protected, it provides an authenticator for the data or message. Any alteration in the original message will produce a very different hash or digest value. The most widely used hash function, called Secure Hash Algorithm-1 (SHA-1), was developed by NIST, to be used with the Digital Signature Algorithm, and was published in 1995 as FIPS 180-1.

Hot plugging - The ability to add and remove devices to a computer while the computer is running and have the operating system automatically recognize the change. This is a feature of PCMCIA, Universal Serial Bus (USB) Standard, and IEEE 1394 Standard FireWire.

I

IA - Information Assurance - Information operations that protect and defend information and information systems by ensuring their confidentiality, authentication, availability, integrity, and non-repudiation.

IA (also) - Intentional (cyber warfare) Attack - Any attack through cyber-means to intentionally affect national security (cyber warfare) or to further operations against national security. Includes cyber attacks by unintentional actors prompted by intentional actors. National policy at the level of warfare.

I-actors - Intentional cyber actors - Individuals intentionally prosecuting cyber warfare (i.e., cyber operators, cyber troops, cyber warriors, cyber forces).

IAFIS - Integrated Automated Fingerprint Identification System - The Federal Bureau of Investigation's Integrated Automated Fingerprint Identification System (IAFIS), implemented in July 1999, replaces a paper-based system for identifying and searching criminal history records. IAFIS supports a law enforcement agency's ability to digitally record fingerprints and electronically exchange information with the FBI. Manual steps have been reduced and processing speeds increased.

IAM - Information Assurance Manager

IASO - Information Assurance Security Officer

IBIA - International Biometric Industry Association - A trade association founded in 1998 to look after the collective international interests of the biometric industry. Governed by and for biometric developers manufacturers, and integrators, and impartially serves all biometric technologies in all applications

IC - Intelligence Community - The United States Intelligence Community is a cooperative federation of sixteen United States government agencies and organizations that work separately and together to conduct intelligence activities considered necessary for the conduct of foreign relations and the protection of the national security of the U.S. The Intelligence Community is led by the Director of National Intelligence. Among their varied responsibilities, the members of the Community collect and produce foreign and domestic intelligence, contribute to military planning, and perform espionage. The Intelligence Community was established by Executive

Order 12333, signed on December 4, 1981 by President Ronald Reagan.

ICANN - Internet Corporation for Assigned Names and Numbers

ICC - Integrated Circuit Card

ICDT - Integrated Capabilities Development Team

ICE - Iris Challenge Evaluation - A large-scale development and independent technology evaluation activity for iris recognition systems sponsored by the U.S. Government in 2005.

ICE (also) - Immigration and Customs Enforcement - Created in March 2003, ICE is the largest investigative branch of the Department of Homeland Security (DHS). The agency was created after 9/11, by combining the law enforcement arms of the former Immigration and Naturalization Service (INS) and the former U.S. Customs Service, to more effectively enforce our immigration and customs laws and to protect the United States against terrorist attacks. ICE does this by targeting illegal immigrants: the people, money and materials that support terrorism and other criminal activities. ICE is a key component of the DHS "layered defense" approach to protecting the nation.

ICR - Information Collection Request

ICT - Information and Communications Technology

IDE - Intrusion Detection Equipment

IDENT/IAFIS - Identification System/Integrated Automated Fingerprint Identification System

IdM - Identity Management

IDMS - Identity Management System - Identifies individuals in a system and controls their access to resources within that system by associating user rights and restrictions with each identified individual. The FIPS 201 standard requires that an identity management system be used to manage the identity information required for the Personal Identity Verification process specified in the standard.

IdP - Identity Provider

IDR - Identification Record Report

IDS - Intrusion Detection System

IDS (also) - Identity Superiority

iDSM - interagency Data Sharing Model

IDS-MD - Identity Dominance System Maritime Domain

IEA - Information Exchange Agreement

IEC - International Electrotechnical Commission

IED - Improvised Explosive Device

IETF - Internet Engineering Task Force

IICT - Interagency Intelligence Committee on Terrorism

III - Interstate Identification Index - System of federal and state criminal history records maintained by the Federal Bureau of Investigation.

IMA - Intelligence Materiel Activity

IMD - Intelligence Material Detachment

IMF - Identity Management Federation

IMINT - Imagery Intelligence

IMP - Impression Type

INCITS - International Committee for Information Technology Standards - Organization that promotes the effective use of information and communication technology through standardization in a way that balances the interests of all stakeholders and increases the global competitiveness of the member organizations. *See also ANSI, ISO, and NIST.*

INFOSEC - Information Systems Security

IN-SAP - Intelligence Special Access Program

INSCOM - U.S. Army Intelligence and Security Command - INSCOM was organized in January 1977 as a result of the Army's Intelligence Organization and Stationing Study (IONS). INSCOM is a relatively new arrival on the Army scene. To understand the roots of the command, it is necessary to go back in time and examine the history of the three main elements which were originally combined to form INSCOM in 1977: the US Army Security Agency

(USASA); U.S. Army Intelligence Agency (USAINTA); and a number of different intelligence production agencies, most of which had been under the Assistant Chief of Staff for Intelligence (ACSI) for direct control.

I&E - Installations and Environment

INSCOM - Intelligence and Security Command

INSPASS - Immigration and Naturalization Service's Passenger Accelerated Service System

IOC - Initial Operating/Operational Capacity

IOC (also) - Intelligence Operations Center

IOSS - Interagency OPSEC Support Staff

IOT&E - Initial Operational Test and Evaluation

IP - Internet Protocol

IPB - Intelligence Preparation of the Battlespace - IPB is an analytical methodology employed to reduce uncertainties concerning the enemy, environment, and terrain for all types of operations. Intelligence preparation of the battlespace builds an extensive database for each potential area in which a unit may be required to operate. The database is then analyzed in detail to determine the impact of the enemy, environment and terrain on operations and presents it in graphic form. Intelligence preparation of the battlespace is a continuing process.

IPMSCG - Identity Protection & Management Senior Coordinating Group

IPR - Intellectual Property Rights

IPT (also) - Interoperability Integrated Project Team

IPT (also) - Integrated Product Team

IR&D - Independent Research and Development

IRTPA - Intelligence Reform and Terrorism Prevention Act (2004)

IS - Information System

ISCAP - Interagency Security Classification Appeals Panel

ISDN - Integrated Services Digital Network

ISE - Information Sharing Environment

ISL - Industrial Security Letter

ISO - International Organization for Standardization - A non-governmental network of the national standards institutes from 151 countries. The ISO acts as a bridging organization in which a consensus can be reached on solutions that meet both the requirements of business and the broader needs of society, such as the needs of stakeholder groups like consumers and users. *See also ANSI, INCITS, and NIST.*

ISOO - Information Security Oversight Office

ISR - Intelligence, Surveillance, and Reconnaissance

ISR (also) - Industrial Security Regulation

ISS - Information Systems Security

ISSM - Information Systems Security Manager

ISSO - Information Systems Security Officer

ISSR - Information Systems Security Representative

IT - Information Technology - IT is concerned with the use of technology in managing and processing information, especially in large organizations. In particular, IT deals with the use of electronic computers and computer software to convert, store, protect, process, transmit, and retrieve information. For that reason, computer professionals are often called IT specialists or Business Process Consultants, and the division of a company or university that deals with software technology is often called the IT department. Other names for the latter are information services (IS) or management information services (MIS), managed service providers (MSP).

ITAR - International Traffic in Arms Regulations, 22 CFR 120-130

ITC - Interagency Training Center

ITF - International Terrorist File

ITL - Information Technology Lab

ITOP - International Test Operations Procedures

ITR - Information Technology Request

ITU-T - International Telecommunication Union - Telecommunication Standardization Sector

IUCRC - Industry/University Cooperative Research Center

IV&V - Independent Verification and Validation

IW - Information warfare - The use and management of information in pursuit of a competitive advantage over an opponent. Information warfare may involve collection of tactical information, assurance that one's own information is valid, spreading of propaganda or disinformation among the enemy, undermining the quality of opposing force information and denial of information collection opportunities to opposing forces.

IWE - Interim Working Environment

Identification - 1) Process by which the biometric system identifies a person by performing a one-to-many (1: n) search against the entire enrolled population. 2) The one-to-many process of comparing a submitted biometric sample against all biometric reference templates on file to determine whether it matches any of the templates and, if so, the identity of the enrollee whose template was matched. 3) The process of using a submitted biometric sample for comparison against a template to match a user to a known enrollee. Normally used only in one-to-many systems. In identification systems, the user makes either no claim or an implicit "negative" claim to an enrolled identity, and a "one-to-many" search of the entire enrolled database is required. 4) A one-to-many or one-to-few process of comparing an individual's biometric sample against a database of biometric reference templates in order to: a) discern the identity of an enrolled individual making an implicit positive claim on identity; or b) ensure the absence of a reference template for an individual making an implicit negative claim on identity. 5) The biometric system using the one-to-many approach is seeking to find an identity amongst a database rather than authenticate a claimed identity. 6) A task where the biometric system searches a database for a reference matching a submitted biometric sample and, if found, returns a corresponding identity. A biometric is collected and compared to all the references in a database. Identification is "closed-set" if the person is known to exist in the database. In "open-set" identification, sometimes referred to as a "watch list," the person is not guaranteed to exist in the database. The system must determine whether the person is in the database, then return the identity. *See also closed-set identification, open-set identification, verification, and watch list.*

Identification algorithm - The procedure of comparing verification data with more than one enrollment data or data set, and calculating the degree of matching score with the enrollment data or data sets.

Identification device - A device comprising an identification algorithm and a face scanner. It generates verification data for each attempt, identifies and ranks all enrollment data or data sets by matching score.

Identification rate - The rate at which an individual in a database is correctly identified.

Identification system - Where the user makes no explicit claim to identity, may be compared to verification systems. Without a claimed identity, the biometric system does a one-to-many process of comparison against all enrollees in its database.

Identifier - A unique data string used as a key in the biometric system to name a person's identity and its associated attributes. An example of an identifier would be a passport number.

Identity - 1) The common sense notion of personal identity, a person's name, personality, physical body, and history, including such attributes as nationality, educational achievements, employer, security clearance, financial and credit history, etc. In a biometric system, identity is typically established when the person is registered in the system through the use of so-called "breeder documents" such as birth certificate, passport, etc. In the context of an access control system, identify means to locate the security system's stored identity information that is associated with the security card or other credential presented to the system, and in some cases performing additional verification using that information such as checking a PIN, comparing a stored biometric to a captured biometric, or performing human visual verification of identifying characteristics. This is called authentication. Where roles are used to assign system privileges, it may be sufficient to securely identify the role of the person rather than the individual personal identity when performing authentication; within the context of a business system or security system, identity generally has one of two meanings. First, it refers to identity information (such as an identifying name or number) that is unique within the system, plus additional information that usually includes one or more of the following: identifying characteristics, which individuals and systems will use to perform an identification; system or organizational role, used to determine the specific rights and authority granted; and the period of time for which the identify information may be relied upon. Oftentimes one particular part of the identity information is referred to as the identity, such as a name or a role within the system. Second, identity can refer to a person, physical object (such as a security smart card), data object (such as a biometric signature on a card) or computer system that is being verified as authentic by the system. 2) Within the context of a Personal Identity Verification system defined by FIPS 201, identify means the real-world process of visually and physically verifying an individual's identity by verifying identification documents and conducting an in-person interview, before registering the person into the identity management system. This initial verification of identity is referred to as identification.

Identity authentication assurance level - There are three identity authentication assurance levels defined in FIPS 201. They express the level of confidence that the cardholder has presented a credential that correctly references the cardholder's identity. The three levels defined are named Some Confidence, High Confidence, and Very High Confidence. The following terms also have the same meaning and are used interchangeably: PIV authentication levels, PIV assurance levels, identity assurance levels, authentication assurance levels and assurance levels.

Identity governance - The combination of policies and actions taken to ensure enterprise-wide consistency, privacy protection and appropriate interoperability between individual identity management systems.

Identity management (IdM) - The combination of systems, rules and procedures that defines an agreement between an individual and organization(s) regarding ownership, utilization, and safeguard of personal identity information.

Image type - An input type in a matching-algorithm and a verification device.

Implicit claim of identity - In applications where there is an implicit claim of identity or non-identity, the submitted sample may need to be matched against many enrolled templates. In this case the accept/reject decision depends on the result of a many comparisons.

Impostor - A person, unknown in the collection of reference templates, who submits his/her own biometric sample in either an intentional or inadvertent attempt to pass him/herself as a legitimate enrollee; A person who submits a biometric sample in either an intentional or inadvertent attempt to pass him/herself off as another person who is an enrollee.

Impostor - A person who submits a biometric sample in either an intentional or inadvertent attempt to claim the identity of another person to a biometric system. *See also attempt.*

Impostor attempt - A single "zero-effort" attempt, by a person unknown to the system to match a stored template.

Impostor claim of identity - Making a false positive claim about identity in the system. A user making a false positive claim about identity in the system. The user falsely claims to be someone else, leading to the comparison of a sample with a non-matching template.

Indifferent user - An individual who knows his/her biometric sample is being collected and does not attempt to help or hinder the collection of the sample. For example, an individual, aware that a camera is being used for face recognition, looks in the general direction of the sensor, neither avoiding nor directly looking at it. *See also cooperative user, non-cooperative user, and uncooperative user.*

Individual - In relation to the Privacy Act, an individual is a citizen of the United States or an alien lawfully admitted for permanent residence.

Individual identification system - A system comprising an identification device and a function for ranking enrolled persons by matching score based on the results obtained by the identification device.

Infrared - Light that lies outside the human visible spectrum at its red (low frequency) end.

In-house test - A test carried out entirely within the environs of the biometric developer, which may or may not involve external user participation.

Initial enrollment - The process of enrolling an individual's biometric data for the first time, such that the alternate means of authentication must be used. *See also enrollment and re-enrollment.*

Intentional cyber warfare attack (IA) - Any attack through cyber-means to intentionally affect national security (cyber warfare) or to further operations against national security. Includes cyber attacks by unintentional actors prompted by intentional actors; can be equated to warfare; National policy at the level of warfare.

Interoperability - Interoperability refers to the ability of a system or a product to work with other systems or products without special effort on the part of the customer. In the context of FIPS 201, it also refers to the ability of different Federal agencies to utilize the same PIV card and PIV management processes so that sufficient trust is established to allow one agency to accept and utilize a PIV card created by another agency.

Iris recognition - A biometric modality that uses an image of the physical structure of an individual's iris for recognition purposes. The iris muscle is the colored portion of the eye surrounding the pupil.

Isotope analysis - Reflects the chemistry of the diet and therefore provides a direct measure of the foods or liquids consumed by an individual. This analysis may be useful in linking the geographical decent of an identity.

J

J2 - Intelligence Directorate - Provides military intelligence to warfighters and force planners in the U.S. Joint Forces Command to support force provision, joint training, experimentation, and integration initiatives.

J8 - Force Structure, Resources, and Assessment Directorate, Joint Staff

JABS - Joint Automated Booking System - JABS automates the collection of fingerprint, photographic, and biographical data during the booking process and provides a mechanism to rapidly and positively identify an individual based on a fingerprint submission to the FBI's Integrated Automated Fingerprint Identification System. The JABS booking submissions also provide real-time updating of the FBI's criminal master files that are available to all Federal, State, and local law enforcement agencies. JABS is a secure medium that enables the automation of the criminal booking process, allowing law enforcement agencies to share criminal data electronically to improve criminal identification response times and avoid duplication of booking data entry.

JAG - Judge Advocate General

JAMS - Joint Adjudication Management System

JBOCB - Joint Biometrics Operational Coordination Board

JBSESC - Joint Biometrics Senior Executive Steering Committee

JBTCB - Joint Biometrics Technical Coordination Board - Technical arm of the biometrics coordination process that develops and coordinates the technical analysis of proposed requirements addressing cost, schedule, and performance.

JC2 - Joint Command and Control

JCA-NID - Joint Coordination Activity - Network Identity

JCAVS - Joint Clearance and Access Verification System

JCD - Joint Capabilities Development

JCD (also) - Joint Capabilities Document

JCIDS - Joint Capabilities Integration and Development System - Defense Department leaders created the system in 2003 as a means of identifying and delivering new capabilities to warfighters around the globe. The JCIDS process fosters interoperability and efficiencies among the services so they can more easily function together as a joint force.

JCOS - Joint and Coalition Operations Support - Provides leadership, management oversight, policy guidance, business process development and improvement, and horizontal integration of activities associated with agile acquisition, with emphasis on processes delivering joint and coalition capabilities to the major Combatant Commanders. The JCOS office collaboratively formulates strategic goals and coordinates agile acquisition alternatives to satisfy validated Defense needs for accelerated technology-based capabilities in the one to four- year delivery window, including coordination with the Joint Staff Capabilities Based Planning (CBP) process and with Combatant Command customers. Coordination and horizontal integration activities include discovering synergies and synchronizing transition programs with AS&C oversight and those throughout the DoD. Special effort is focused on addressing joint, coalition, and transformational warfighter needs. JCOS aims to facilitate efficient, timely transition of emergent technologies from Defense, commercial and academic providers—domestic and foreign—into relevant joint and coalition capabilities.

JCS - Joint Chiefs of Staff

JCTD - Joint Capability/Concept Technology Demonstrations

JEFF - Joint Expeditionary Forensic Facility

JEF - Joint Expeditionary Forensics

JFAIDD - Joint Federal Agencies Intelligence DNA Database

JFCOM - U.S. Joint Forces Command

JIC - Joint Intelligence Committee

JIEDDO - Joint Improvised Explosive Device Defeat Organization - JIEDDO leads the U.S. counter-IED effort.

JITC - Joint Interoperability Test Command

JMD - Joint Manning Document

JMIC - Joint Military Intelligence College

JMIP - Joint Military Intelligence Programs

JMITC - Joint Military Intelligence Training Center

JOC - Joint Operating Concept

JPAS - Joint Personnel Adjudication System

JROC - Joint Requirements Oversight Counsel - The JROC reviews programs designated as JROC interest and supports the acquisition review process. In accordance with the CJCS Instruction 3170.01, the Joint Staff reviews all Joint Capabilities Integration and Development System documents and assigns a Joint Potential Designator. The JROC charters Functional Capabilities Boards. The boards are chaired by a JROC-designated chair and, for appropriate topics, co-chaired by a representative of the Milestone Decision Authority. Functional Capabilities Boards are the lead coordinating bodies to ensure that the joint force is best served throughout the Joint Capabilities Integration and Development System and acquisition processes. The Joint Capabilities Integration and Development System process encourages early and continuous collaboration with the acquisition community to ensure that new capabilities are conceived and developed in the joint warfighting context. The JROC, at its discretion, may review any Joint Capabilities Integration and Development System issues which may have joint interest or impact. The JROC will also review programs at the request of, and make recommendations as appropriate to, the Secretary of Defense, Deputy Secretary of Defense, Under Secretary of Defense (Acquisition, Technology, and Logistics), Assistant Secretary of Defense (Networks and Information Integration), and the Under Secretary of the Air Force (as DoD Space Milestone Decision Authority). The JROC also validates key performance parameters.

JROCM - Joint Requirements Oversight Council Memorandum

JRWG - Joint Requirements Working Group

JSIDS - Joint Capabilities Integration and Development System

J-SIIDS - Joint Services Interior Intrusion Detection System

JSOC - Joint Special Operations Command

JSP - Joint Service Program

J-TIDS - Joint Tactical Information Distribution System

JUONS - Joint Urgent Operational Needs

JWICS - Joint World-Wide Intelligence Communications System

K

K - Number of partitions in a filtering or binning method

KMP - Key Management Personnel

KST - Known and Suspected Terrorist

Key - In encryption and digital signature a string of bits used for encrypting and decrypting information to be transmitted. Encryption commonly relies on two different types of keys, a public key and a private key.

Keystroke dynamics - A biometric modality that uses the cadence of an individual's typing pattern for recognition.

Key management - The various processes that deal with the creation, distribution, authentication, and storage of keys.

Keystroke recognition - Keystroke recognition analyzes the way an individual types. Users enroll in a system by typing the same word or words several times. The system verifies the user by recognizing the distinctive rhythm a person uses while typing.

Key reference - A PIV key reference is a one-byte identifier that specifies a cryptographic key according to its PIV Key Type. The identifier is part of cryptographic material used in a cryptographic protocol such as an authentication or a signing protocol.

L

L&O - Law and Order

LAA - Limited Access Authorization

LAC - Local Agency Check

LACS - Logical Access Control Systems

LAN - Local Area Network

LC - Joint Staff Office of Legal Counsel

LCR - Listed Character Reference

LEP - Local Employed Person

LFA - Local Feature Analysis - Pertains to matching facial regions defined by features such as the nose, eyes, and mouth, rather than the face as a whole, by making explicit use of *a priori* knowledge about the structure of the human face.

LFC - Local Files Check

LIMDIS - Limited Dissemination

LOC - Letter of Consent

LOD - Letter of Denial

LOI - Letter of Intent

LON - Letter of Notification

LSB - Least Significant Bit

Latent fingerprint - 1) Transferred impression of friction ridge detail not readily visible. 2) Generic term used for questioned friction ridge detail. 3) A fingerprint image left on a surface that was touched by an individual. The transferred impression is left by the surface contact with the friction ridges, usually caused by the oily residues produced by the sweat glands in the finger. *See also friction ridge.*

Live capture - 1) The process of capturing a biometric sample by an interaction between an end user and a biometric system. 2) Typically refers to a fingerprint capture device that electronically captures fingerprint images using a sensor (rather than scanning ink-based fingerprint images on a card or lifting a latent fingerprint from a surface). *See also sensor.*

Live scan - Occurs when taking a fingerprint or palm print directly from a subject's hand.

Liveness detection - A technique used to ensure that the biometric sample submitted is from an end user. A liveness detection method can help protect the system against some types of spoofing attacks. *See also challenge response, mimic, and spoofing.*

Logical credential - *See electronic credential.*

Loop - A fingerprint pattern in which the friction ridges enter from either side, curve sharply and pass out near the same side they entered. This pattern will contain one core and one delta. *See also arch, core point, delta point, friction ridge, and whorl.*

M

M - Number of samples submitted during each transaction.

M1 - The U.S. National Standards development organization, run under the auspices of INCITS.

m - Number of samples used in an initial search.

M&S - Modeling and Simulation

MAA - Mission Area Analysis

MAC - Mandatory Access Control

MAC - Military Airlift Command

MACL - Mission Area Capability Library

MACOM - Major Command

MAFIS - Maritime Automated Fingerprint ID System ID Book

MAIS - Major Automated Information Systems

MAJCOM - Major Command

MAM - Military-Aged Male - Reference to enemy forces.

MANPRINT - Manpower and Personnel Integration

MASINT - Measurement and Signature Intelligence

MC - Mitigating Conditions

MCCDC - U.S. Marine Corps, Combatant Development Command

MCTL - Militarily Critical Technologies List

MDA - Maritime Domain Awareness

MDA (also) - Missile Defense Agency

MDAP - Major Defense Acquisition Program

MEF - Mission Essential Functions

METL - Mission Essential Task List

MFO - Multiple Facility Organization

MFT - Multi-Function Teams

MI - Military Intelligence

MILCOM - Military Communication

MIL-STD - Military Standard

MIME - Multipurpose Internet Mail Extensions

MISWG - Multinational Industrial Security Working Group

MNC-I - Multi National Corps Iraq

MNF-I - Multi-National Force - Iraq

MNF-I access control policy - This system is a response to an urgent requirement to better secure U.S. and

Coalition bases from MNF-I. It sets a universal policy to collect biometrics, screen, and then issue an access badge to Iraq and third-country nations needing access to Coalition bases. The card issued is designed to be a vast improvement over the multitude of cards currently in use across the Iraqi Theater of Operations. This system is being phased out as BISA matures.

MNS - Mission Needs Statement

MO - Method of Operation

MOA - Memorandum of Agreement

MOBS - Mission Oriented Biometric Software

MOS - Military Occupational Specialty

MOTS - Modified/Modifiable/Military Off-the-Shelf - A MOTS product is typically a COTS product whose source code can be modified. The product may be customized by the purchaser, by the vendor, or by another party to meet the requirements of the customer. In the military context, MOTS refers to an off-the-shelf product that is developed or customized by a commercial vendor to respond to specific military requirements. Because a MOTS product is adapted for a specific purpose, it can be purchased and used immediately. However, since MOTS software specifications are written by external sources, government agencies are sometimes leery of these products, because they fear that future changes to the product will not be in their control.

MOU - Memorandum of Understanding

MP - Military Police

MPAC - Multi-Purpose Access Card - MPAC is operated by Iraqi interim government (IIG) to process Iraqi citizens for the new Iraqi Government. The current focus of collection is on employees of the IIG. However, there are plans to expand the system to encompass the entire adult population.

MRTD - Machine Readable Travel Document

MRT - Machine Readable Technology

MSB - Most Significant Bit

MSIC - Military and Space Intelligence Center

MTAC - Multiple Threat Alert Center - The Department of the Navy's MTAC provides indications and warning for a wide range of threats to Navy and Marine Corps personnel and assets around the world. Operated by the Naval Criminal Investigative Service (NCIS), the MTAC utilizes NCIS' worldwide presence and combination of law enforcement, counterintelligence, intelligence and security capabilities to identify all available threat indicators. Analysts, special agents, and military personnel work in the MTAC around the clock to produce indications and warning of possible terrorist activity, foreign intelligence threats and criminal threats that may affect naval operations. The MTAC is an outgrowth of the Navy Antiterrorist Alert Center (ATAC). The ATAC was established in December 1983 to address the terrorist threat after the bombing of the Marine Corps barracks in Beirut, Lebanon, and the murders of Navy officers in Greece and Central America. As the first 24-hour terrorism watch center in the U.S. intelligence community, the ATAC successfully supported the Navy and Marine Corps team for nearly two decades.

mtDNA - mitochondrial DNA

MTMC - Military Traffic Management Command

ManTech - Manufacturing Technology

Match - A decision that a biometric sample and a stored template comes from the same human source, based on their high level of similarity (difference or hamming distance). *See also false match rate, and false non-match rate.*

Match - 1) The process of comparing a biometric sample against a previously stored template and scoring the level of similarity. When comparing a score to a biometric system threshold, a positive result is considered a match, while a negative result is considered a non-match. 2) The process of computing a score (of similarity) between a biometric sample and a given reference template.

Match-evaluating function - To determine whether a successful match is obtained (an authorized individual) or not obtained (an unauthorized individual) based on the output from the verification device in an authentication system.

Matching - 1) The act of comparing a piece of verification data against a piece of enrollment data and calculating the matching score or distance between them. 2) The process of comparing a biometric sample against a previously stored template and scoring the level of similarity. A accept or reject decision is then based upon whether this score exceeds the given threshold. 3) The process of comparing a biometric sample against a previously stored template and scoring the level of similarity. Systems then make decisions based on this score and its relationship (above or below) a predetermined threshold. 4) The comparison of biometric templates to determine their degree of similarity or correlation. A match attempt results in a score that, in most systems, is compared against a threshold. If the score exceeds the threshold, the result is a match. If the score falls below the threshold, the result is a non-match. The purpose of matching is to obtain some measure of the degree of match/similarity between the face patterns of an unknown subject and an enrolled subject. *See also comparison, difference score, and threshold.*

Matching algorithm - 1) An algorithm used to verify one item of verification data against one item of enrollment data to determine matching score or distance. 2) The algorithm used to compare a piece of enrollment data against an iris image and to calculate the matching score or distance between them.

Matching decision - 1) To compare one item of verification data with one item of enrollment data (or compare a verification data set with a enrollment data set) and to determine whether they are matched. 2) The act of determining a match or non-match by comparing a piece of verification data against a piece of enrollment data, or comparing a verification data set against an enrollment data set.

Matching determination - The act of determining a match or non-match by comparing verification data with enrollment data, or comparing a verification data set with an enrollment data set. The enrollment data of another person may be referred to during matching determination.

Matching score - 1) A measure of similarity or dissimilarity between the biometric data and a stored template, used in the comparison process. 2) A value representing the matching score between verification data and enrollment data. The larger the value the more similar the data. Dissimilarity or "distance" can be used as an index representing the similarity between verification and enrollment. Not only the enrollment data of a person but also enrollment data or enrollment data sets of other persons may be used in the calculation of matching score.

Mate - Another one when computed from two samples captured from the same physical trait of the same (e.g., two different impressions of the same fingerprint).

Mimic - The presentation of a live biometric measure in an attempt to fraudulently impersonate someone other than the submitter. *See also challenge response, liveness detection, and spoofing.*

Minutia(e) point - Friction ridge characteristics that are used to individualize a fingerprint image, see illustration below. Minutiae are the points where friction ridges begin, terminate, or split into two or more ridges. In many fingerprint systems, the minutiae (as opposed to the images) are compared for recognition purposes. *See also friction ridge, and ridge ending.*

Minutiae - The point where a friction ridge begins, terminates, or splits into two or more ridges. Minutiae are friction ridge characteristics that are used to individualize a fingerprint image.

Mitochondrial DNA analysis - mtDNA is maternally inherited, so that all family members who share a direct maternal lineage share the same mtDNA profile. MtDNA is an excellent tool for kinship analyses. However, profiles generated from mtDNA usually cannot be used to uniquely identify an individual.

Modality - A type or class of biometric system. Examples include face recognition, fingerprint recognition, iris recognition, etc.

Model - A representation used to characterize an individual. Behavioral-based biometric systems, because of the inherently dynamic characteristics, use models rather than static templates. *See also template.*

Multi-biometric authentication - Authentication using two or more different biometric types.

Multi-factor authentication - Authentication using two or more identity factors: 1) knowledge factor—something an individual knows; 2) possession factor—something an individual has; and 3) biometric factor—something an individual is.

Multimodal biometric - A biometric device which uses information from different biometrics (e.g., fingerprint and hand shape or fingerprints from two separate fingers).

Multimodal biometric system - A biometric system in which two or more of the modality components (biometric characteristic, sensor type or feature extraction algorithm) occurs in multiple.

Multimodal system - 1) A biometric system that includes more than one biometric system or biometric technology. 2) Biometric system that uses two or more different biometric characteristics or technologies (i.e., two fingerprint algorithms; fingerprint + iris; iris + face; etc.)

Multiple biometric - A biometric system that includes more than one biometric system or biometric technology. *See also multimodal system.*

Multi-value type - An output type in a matching algorithm and a verification device, to output the matching score or distance.

Mutual authentication - Mutual authentication or two-way authentication refers to two parties authenticating each other suitably. In technology terms, it refers to a requesting/asserting entity authenticating themselves to a relying party/server and that server authenticating itself to the requesting/asserting entity in such a way that both parties are assured of the others' identity. There is also the less trustworthy one-way authentication. This is where only the requesting/asserting entity undergoes authentication to the relying party/server, so the requesting/asserting entity never gets to verify that they are communicating with the correct relying party/server.

N

N - Number of active stored templates or models in the database.

NAC - National Agency Check - A NAC consists of a check of the files of a number of government agencies for pertinent facts bearing on the loyalty and trustworthiness of the   individual. Examples of agencies checked are the FBI and the Defense Central Index of Investigations.

NACI - National Agency Check plus Written Inquiries

NACLC - National Agency Check/Local Agency Check/ Credit Check

NAF - Non-appropriated Funds

NAIL - Naval Innovation Lab

NARA - National Archives and Records Administration

NASA - National Aeronautics and Space Administration

NATIBO - North American Technology and Industrial Base Organization - The North American Technology and Industrial Base Organization is chartered to promote a cost effective, healthy technology and industrial base that is responsive to the national and economic security needs of the United States and Canada.

NATO - North Atlantic Treaty Organization

NBTC - National Biometric Test Center

(NC) - NATO Confidential

NC - NOCONTRACT

(NCA) - NATO Confidential ATOMAL

NCA - National Command Authority

NCAF - Department of Navy Central Adjudication Facility

NCHC - National Criminal History Check

NCIC - National Crime Information Center

NCIS - Naval Criminal Investigative Service

NCISSE - National Colloquium Information Systems Security Education

NCIX - National Counter Intelligence Executive

NCMS - National Classification Management Society

NCR - National Capital Region

NCS - National Cryptologic School

NCSC - National Computer Security Center

NCSC (also) - National Communications Security Committee (replaced by NSTISSC)

NCTC - National Counterterrorism Center - In 2004, President Bush established the NCTC to serve as the primary organization in the United States Government (USG) for integrating and analyzing all intelligence pertaining to terrorism and counterterrorism (CT) and to conduct strategic operational planning by integrating all instruments of national power. In December 2004, Congress codified the NCTC in the Intelligence Reform and Terrorism Prevention Act (IRTPA) and placed the NCTC in the Office of the Director of National Intelligence. Located at the Liberty Crossing Building in Northern Virginia, the NCTC is a multi-agency organization dedicated to eliminating the terrorist threat to U.S. interests at home and abroad.

NCW - Network-Centric Warfare - Now commonly called Network-centric operations (NCO), is a new military doctrine or theory of war pioneered by the United States Department of Defense. NCW/NCO is an emerging theory of war in the information age that seeks to translate an information advantage into a competitive warfighting advantage through the robust networking of well informed geographically dispersed forces allowing new forms of organizational behavior. This "networking" utilizes information technology via a robust network to allow increased information sharing, collaboration, and shared situational awareness, which, theoretically allows greater self-synchronization, speed of command, and mission effectiveness. The theory hypothesis has four basic tenets: 1) increased mission effectiveness; 2) a robustly networked force improves information sharing; 3) information sharing enhances the quality of information and shared situational awareness; and 4) shared situational awareness enables collaboration and self-synchronization and enhances sustainability and speed of command. Network Enabled Capability (NEC) is a term used in the United Kingdom and elsewhere for a similar doctrine.

NDA - Non-Disclosure Agreement

NDI - Non-Developmental Items

NDIA - National Defense Industrial Association - Provides individuals from academia, government, the military services, small businesses, prime contractors, and the international community, the opportunity to network effectively with the government-industry team, keep abreast of the latest in technology developments, and address and influence issues as well as government policies critical to the health of the defense industry and the preservation of our national security.

NDP - National Disclosure Policy

NDP-1 - National Policy and Procedures for the Disclosure of Classified Military Information to Foreign Governments and International Organizations

NDPC - National Disclosure Policy Committee

NDRR - Non-Identification Response

NF - *See NOFORN.*

N-FACS - National Fingerprint based Applicant Check Study

NFIP - National Foreign Intelligence Program

NGA - National Geospatial-Intelligence Agency - The National Geospatial-Intelligence Agency (NGA) is an agency of the United States Government with the primary mission of collection, analysis, and distribution of geospatial intelligence (GEOINT) in support of national security. NGA was formerly known as the National Imagery and Mapping Agency (NIMA) and is part of the Department of Defense (DoD). In addition, NGA is a member agency of the United States Intelligence Community. NGA's headquarters are located in Bethesda, Maryland and operates major facilities in the Northern Virginia, Washington, D.C., and St. Louis, Missouri areas as well as support and liaison offices worldwide. In 2011 NGA expects to consolidate many of its Washington DC, Maryland, and Northern Virginia activities in a new east-coast campus near Ft. Belvoir as part of the BRAC. Its budget and number of employees are classified.

NGA (also) - Next Generation ABIS

NGIC - National Ground Intelligence Center

NGN - Next Generation Network

NIAP - National Information Assurance Partnership - Collaboration between the National Institute of Standards and Technology (NIST) and the National Security Agency (NSA) that provides security testing, evaluation, and validation programs in accordance with the Common Criteria standard.

NIJ - National Institute of Justice

NIMA - National Imagery and Mapping Agency - Currently known as the National Geospatial-Intelligence Agency, formerly Defense Mapping Agency, formerly Army Mapping Service. *See NGA.*

NIPC - National Infrastructure Protection Center

NISP - National Industrial Security Program

NISPOM - National Industrial Security Program Operating Manual

NISPOMSUP - National Industrial Security Program Operating Manual Supplement

NIST - National Institute of Standards and Technology - U.S. governmental standards group that publishes the Federal Information Processing Standards, including FIPS 201. A non-regulatory federal agency within the U.S. Department of Commerce that develops and promotes measurement, standards, and technology to enhance productivity, facilitate trade, and improve the quality of life. NIST's measurement and standards work promotes the well-being of the nation and helps improve, among many others things, the nation's homeland security. *See also ANSI, INCITS, and ISO.*

NIT WOT - National Implementation Plan War on Terror

Nlets - The International Justice and Public Safety Information Sharing Network (previously known as NLETS).

NMIC - National Maritime Intelligence Center

NMIST - National Military Intelligence Support Terminal

NN - Nickname

NOCONTRACT - Not Releasable to Contractors or Contractor Consultants

NOFORN - Not Releasable to Foreign Nationals

NOL - NCTC Online - Serves as the counterterrorism community's library of terrorism information. This repository reaches the full range of intelligence, law enforcement, military, homeland security, and other Federal organizations involved in the global war on terrorism.

(NR) - NATO Restricted

NRO - National Reconnaissance Office

(NS) - NATO Secret

(NSA) - NATO Secret ATOMAL

NSA - National Security Agency - NSA is the U.S. government's Cryptologic organization. Officially established on November 4, 1952, it is believed to be the world's largest intelligence-gathering agency. Responsible for the collection and analysis of foreign communications, if coordinates, directs, and performs highly specialized activities to produce foreign signals intelligence information, which involves a significant amount of cryptanalysis. It is also responsible for protecting U.S. government communications from similar agencies elsewhere, which involves a significant amount of cryptography. A component of the Department of Defense, the NSA is a key component of the United States Intelligence Community headed by the Director of National Intelligence.

NSC - National Security Council

NSCTT- National Strategy to Combat Terrorist Travel

NSD - National Security Directive

NSDD - National Security Decision Directive

NSDM - National Security Decision Memorandum

NSI - National Security Information

NSM - Network Security Manager

NSTISSC - National Security Telecommunications and Information Systems Security Committee (#11) - National security community policy governing the acquisition of information assurance (IA) and IA-enabled information technology products.

NTC - National Targeting Center

NTISSI - National Telecommunications and Information Systems Security Instruction

NTK - Need To Know

(NU) - NATO Unclassified

NW3C - National White Collar Crime Center

National identity card schemes - Many Countries are developing government certified national identity card schemes based on PKI certificates deployed in smartcards, mobile SIM solutions or soft certificates. The list of initiatives is very long, but information about European Initiatives can be found at this URL. Many countries are encouraging public-private partnerships using the eID cards within commercial applications. This has advantages for large-scale roll-out of strong authentication mechanisms since many commercial organizations are reluctant to underwrite the costs and liabilities involved in issuing tokens.

Negative claim - A claim by a user not to be enrolled in the biometric system. This may be needed to establish that double claims are not being made.

Negative claim of identity - The claims (either implicitly or explicitly) not to be known to or enrolled in the system. Enrollment in social service systems open only to those not already enrolled is an example.

Negative identification (background check) - Negative identification is the process of verifying that a subject's biometric characteristics are not contained in a large database of previously enrolled persons.

Neural net/network - 1) One particular type of algorithm. An artificial neural network uses artificial intelligence to learn by past experience and compute whether a biometric sample and template are a match. 2) A type of algorithm that learns from past experience to make decisions *See also algorithm.*

Noise - Unwanted components in a signal that degrade the quality of data or interfere with the desired signals processed by a system.

Non-cooperative user - An individual who is not aware that his/her biometric sample is being collected, Example: A traveler passing through a security line at an airport is unaware that a camera is capturing his/her face image. *See also cooperative user, indifferent user, and uncooperative user.*

Non-repudiation - Assurance that the sender is provided with proof of delivery and that the recipient is provided with proof of the sender's identity so that neither can later deny having processed the data.

Normalization - Regarding face patterns, refers to removing any variation that does not pertain to identity.

O

ô - Similarity or distance score threshold.

O&M - Operations and Maintenance

OADR - Originating Agency's Determination Required

OASIS - Organization for the Advancement of Structured Information Standards

OC - *See ORCON.*

OCA - Original Classification Authority

OCONUS - Outside of the Continental United States

OCSP - Online Certificate Status Protocol - A method for systems to verify the status of a digital certificate (to determine whether or not it has been revoked) by sending a status query to a server and receiving a real-time response about the status of the certificate.

ODC - Office of Defense Cooperation

OEA - Office of Economic Adjustment

OEF - Operation Enduring Freedom

OEM - Original Equipment Manufacturer - In the context of biometrics, a manufacturer that assembles a complete biometric system from parts.

OEM (also) - Original Equipment Module - A biometric module for integration into a complete biometric system

ØG (D) - "Genuine" distance distribution function.

OGC - Office of General Counsel

ØI (D) - "Impostor" distance distribution function.

OI - Operating Instruction

OID - Object Identifier

OIF - Operation Iraqi Freedom

OMB - Office of Management and Budget

OPF - Official Personnel File

OPLOC - Operating Location

OPM - Office of Personnel Management

OPSEC - Operations Security

OR - Operations Research

ORCON - Dissemination and Extraction of Information Controlled by Originator

ORD - Operational Requirements Document

ORI - Originating Agency Identifier

OSD - Office of the Secretary of Defense - OSD is part of the United States Department of Defense and includes the entire staff of the Secretary of Defense. It is the principal staff element of the Secretary of Defense in the exercise of policy development, planning, resource management, fiscal, and program evaluation responsibilities. OSD includes the immediate offices of the Secretary and Deputy Secretary of Defense, as well as five Under Secretaries in the fields of Acquisition, Technology & Logistics; Comptroller/Chief Financial Officer; Intelligence; Personnel & Readiness; and Policy. All of these positions require U.S.

Senate confirmation, and each has its own staff. Other positions include the Director of Defense Research and Engineering, Assistant Secretaries of Defense, General Counsel, Director of Operational Test and Evaluation, Assistants to the Secretary of Defense, Director of Administration and Management, and such other staff offices as the Secretary establishes to assist in carrying out assigned responsibilities.

OSD AT&L I&E - Office of the Secretary of Defense Acquisition Technology and Logistics Inspection and Evaluation

OSD-ATL - Office of Secretary of Defense Acquisition Technology and Logistics

OSI - (Air Force) Office of Special Investigations

OSTP - Office of Science & Technology Policy

ØT (D) - Inter-template distance distribution function

OTA - Office of Technical Assessment

OUID - Organization Unique Identifier - An OUID is a unique, simple, and non-intelligent identifier that will be used to uniquely identify Department of Defense (DoD) and non-DoD organizations, including, but not limited to, U.S. and foreign federal, civil and commercial entities. The OUID will be a subset of the Force Management Identifiers (FMIDS) that will be used to identify doctrinal organizational information in the Global Force Management Information Exchange Data Model (GFMIEDM) format as maintained in the Global Force

Management Organization Servers. The GFMIEDM in the Organization Servers will provide a means for the entire DoD Enterprise to maintain and exchange force structure information in support of War Fighter and Business Mission Area operations. The OUID is maintained by the OUID Registry.

One-to-few - A hybrid of one-to-many identification and one-to-one verification, typically, the one-to-a-few process involves comparing a submitted biometric sample against a small number of biometric reference templates on file. Similar to one-to-many but with a small (or a subset of the) database, this could imply comparing a biometric sample to reference templates from different individuals, or from a single individual.

One-to-many - 1) Refers to the comparison of an individual's biometric sample against all reference templates in a given database. This term applies to identification type systems. 2) A phrase used in the biometrics community to describe a system that compares one reference to many enrolled references to make a decision. The phrase typically refers to the identification or watch list tasks. Synonym for identification.

One-to-many matching - *See identification system.*

One-to-one - A phrase used in the biometrics community to describe a system that compares one reference to one enrolled reference to make a decision. The phrase typically refers to the verification task (though not all verification tasks are truly one-to-one) and the identification task can be accomplished by a series of one-to-one comparisons.

Refers to the comparison of an individual's biometric sample against a single reference template. This term applies to a verification-type system. Synonym for verification.

One-to-one matching - *See verification system.*

Ongoing authentication - Confidence in a claimed identity is an ongoing authentication process depending on the assurance levels required. Once credentials based on a claimed identity are issued, whenever they are used, there may be the requirement of determining that the claimed identity is used by the requesting/asserting entity to which the identity was assigned. This ongoing authentication process is needed to minimize the likelihood of fraud based on one entity pretending to be another (sometimes known as identity theft or "phishing").

OpenID - Open Identity - A decentralized single sign-on system. Using OpenID-enabled sites, web users do not need to remember traditional authentication tokens such as username and password. Instead, they only need to be previously registered on a website with an OpenID "identity provider" (IdP). Since OpenID is decentralized, any website can employ OpenID software as a way for users to sign in. OpenID solves the problem without relying on any centralized website to confirm digital identity.

Open-set identification - Biometric task that more closely follows operational biometric system conditions to: 1) determine if someone is in a database; and 2) find the record of the individual in the database This is sometimes referred to as the "watch list" task to differentiate it from the more commonly referenced closed-set identification. *See also closed-set identification, and identification.*

Open-set identification - Identification, when it is possible that the individual is not enrolled in the biometric system. Opposite of closed-set identification.

Open versus Closed - Asks whether the system will be required, now or in the future, to exchange data with other biometric systems run by other management. For instance, some State social service agencies want to be able to exchange biometric information with other States. If a system is to be open, data collection, compression, and format standards are required. This list is open, meaning that additional partitions might also be appropriate. It could also be argued that not all possible partition permutations are equally likely or even permissible.

Operational evaluation - One of the three types of performance evaluations. The primary goal of an operational evaluation is to determine the workflow impact seen by the addition of a biometric system. *See also technology evaluation and scenario evaluation.*

Out of set - In open-set identification, when the individual is not enrolled in the biometric system.

Overt - Biometric sample collection where end users know they are being collected and at what location. An example of an overt environment is the US-VISIT program where non-U.S. citizens entering the United States submit their fingerprint data. *See also covert.*

Overt versus Covert - If the user is aware that a biometric identifier is being measured, the use is overt. If unaware, the use is covert. Almost all conceivable access control and non-forensic applications are overt. Forensic applications can be covert. We could argue that this second partition dominates the first in that a wolf cannot cooperate or non-cooperate unless the application is overt.

P

p *i* - Probability that a sample will be in the *i*th partition.

PA - Privacy Act

PACOM – U.S. Pacific Command

PACS - Physical Access Control System

PACS 2.2 - The short name for a document titled, Technical Implementation Guidance: Smart Card Enabled Physical Access Control Systems, Version 2.2, published in July of 2004 by the Physical Access Interagency Interoperability Working Group (PAIIWG) of the Government Smart Card Interagency Advisory Board (GSC-IAB). The document is also commonly referred to as PACS Implementation Guidance Version. 2.2.

PACS assurance profile - The PACS 2.2 document introduced the term "assurance profiles" and defined high, medium and low assurance profiles. These are similar to but different from the FIPS 201 PIV Identity Authentication Assurance Levels. Some documents refer to these assurance profiles as "PACS assurance levels" or "card assurance levels."

PAO - Public Affairs Officer

PAR - Program Access Request

PBD - Program Budget Decision

PCCF - (Army) Personnel Central Security Clearance Facility

PCL - Personnel Security Clearance

PCMCIA card - A hardware device that supports specific dedicated functions. Examples of PCMCIA card functions include memory devices, input/output devices (e.g., modems and fax modems), and portable disk drives. PCMCIA cards are most commonly used to provide additional computing features for portable computers such as laptops.

PCO - Procurement Contracting Officer

PCU - Premise Control Unit

PDD - Presidential Decision Directive

PDES - Person Data Exchange Standard - A specification of the U.S. government intelligence community that specifies XML tagging of person data including biometric data.

PDM - Program Decision Memorandum

PDR - Preliminary Design Review

PDR (also) - Person Data Repository

PDS - Protected Distribution System

PE - Program Element

PEM - Program Element Monitor

PEO - Program Executive Officer

PEO EIS - Program Executive Office Enterprise Information Systems

PERSEREC - Personnel Security Research Center

PFPA - Pentagon Force Protection Agency

Pi - Penetration rate owing to the ith filter or binning method.

PIA - Privacy Impact Assessment - An analysis of how information is handled: to ensure handling conforms to applicable legal, regulatory, and policy requirements regarding privacy; to determine the risks and effects of collecting, maintaining, and disseminating personally identifiable information in an electronic information system; and to examine and evaluate protections and alternative processes for handling information to mitigate potential privacy risks.

PIC - Personnel Investigations Center (DSS)

PII - Personally Identifiable Information - Any information that permits the identity of an individual to be directly or indirectly inferred, including any other information that is linked or linkable to that individual regardless of whether the individual is a U.S. citizen, legal permanent resident, or a visitor to the U.S.

PIN - Personal Identification Number - A security method used to show "what you know." Depending on the system, a PIN could be used to either claim or verify a claimed identity.

PIV - Personal Identity Verification - Term designated in FIPS 201 for the processes and technologies involved in (a) identification: verifying the identity of a Federal employee or contractor at the time of initial identification and enrollment into a Federal agency's identity management system, and (b) authentication: verifying the identity of the employee or contractor for purposes of physical and information systems access control.

PIV card - A smart card that is designed, issued, and managed according to the specifications in FIPS 201 and its related technical documents.

PIV key type - A type of a key. The PIV key types are: 1) PIV authentication key; 2) PIV card authentication key; 3) PIV digital signature key; 4) PIV key management key; and 5) card application administration key.

PIX - Proprietary Identifier eXtension

PKC - Public Key Cryptography - Encryption system using a linked pair of keys, What one key encrypts, the other key decrypts.

PKCS - Public Key Cryptography Standard

PKE - Public Key Enabled

PKI - Public Key Infrastructure - Portion of the security management infrastructure dedicated to the management of keys and certificates used by public key-based security services. A PKI is a credentials service; it associates user and entity identities with public keys. A well-run PKI is the foundation on which the trustworthiness of public key-based security mechanisms rests.

PKITS - Public Key Infrastructure Test Suite

PM - Program/Program Manager

PM ISE - Program Manager for the Information Sharing Environment - Tasked by Congress to improve terrorism information sharing among Federal and non-Federal entities. Federal agencies, in cooperation with the PM ISE are working to integrate business processes to ensure reliable information flow among Federal, state, local, tribal, and private sector entities.

PMO - Provost Marshal Office

POA - Plan of Action

POA&M - Plan of Action and Milestones

POC - Point of Contact

POM - Program Objective Memorandum

POR - Program of Record

PP - Protection Profile - A form of generic security target defined in the common criteria. Large scale identification (i.e., "Is this person in the database?") Given an input biometric sample, a large-scale identification determines if the pattern is associated with any of a large number (e.g., millions) of enrolled identities. Typical large-scale identification applications include welfare-disbursement, national ID cards, border control, voter ID cards, driver's license, criminal investigation, corpse identification, parenthood determination, missing children identification, etc. These large-scale identification applications require a large sustainable throughput with as little human supervision as possible.

PPBE - Planning, Programming, Budgeting, and Execution

PPBES - Planning, Programming, Budgeting, and Execution System

PPBERS - Planning, Programming and Budgeting Execution Review System

PPBS - Planning, Programming and Budgeting System

Ppi - Pixels per inch

PPII - Protection of Personally Identifiable Information

Ppmm - Pixels per millimeter

(PR) - *See PROPIN.*

PR - Periodic Reinvestigation

PROPIN - Caution, Proprietary Information Involved

PRP - Personnel Reliability Program

PSA - Principal Staff Assistant

PSAB - Personnel Security Appeals Board

P-SAP - Prospective Special Access Program

PSE - Physical Security Equipment - A generic term encompassing any item, device, or system that is used primarily for the protection of Government property, including nuclear, chemical, and other munitions, personnel, installations, and in the safeguarding of national security information and material, including the destruction of such information and material both by routine means and by emergency destruct methods.

PSEAG - Physical Security Equipment Action Group

PSEL - Physical Security Equipment Listing

PSG - Program Security Guide

PSI - Personnel Security Investigation

PSI (also) - Program Security Instruction

PSM - Program Security Manager

PSO - Program Security Officer

PSP - Personnel Security Program

PSQ - Personnel Security Questionnaire

PSS - Protective Security Service

Psys - System Penetration Rate

PGW - Persian Gulf War

PUC - Person Under Control

Palm - A physical biometric that analyses the palm of the hand. Typically this will involve an analysis of minutiae data.

Palm print recognition - A biometric modality that uses the physical structure of an individual's palm print for recognition purposes.

Passive attack - When an impostor submits a biometric sample (and possibly a claim of identity) in an attempt to initiate a false match.

Passive impostor acceptance - When an impostor submits his/her own biometric sample and claiming the identity of another person (either intentionally or inadvertently) he/she is incorrectly identified or verified by a biometric system. Occurs when a person submits his/her own natural biometric sample, attempts to relate it to another individual, and is recognized or verified as that individual. *See also active impostor acceptance.*

Password - Security measure used to restrict access to computer systems and sensitive files. A password is a unique string of characters that a user types in as an identification code. The system compares the code against a stored list of authorized passwords and users. If the code is legitimate, the system allows the user access at whatever security level has been approved for the owner of that password.

Patriot Act - The Uniting and Strengthening America by Providing Appropriate Tools Required to Intercept and Obstruct Terrorism Act of 2001 (Public Law 107-56), known as USA PATRIOT Act or simply the "Patriot Act," is an American act which was signed into law by President George W. Bush on October 26, 2001. The Act passed in the Senate by a vote of 98 to 1, and in the House by a vote of 357 to 66. Although the bill enjoyed widespread Congressional and Presidential support it is a very controversial federal legislation. Originally passed after the September 11, 2001 attacks, the Act was formed in response to the terrorist attacks against the United States, and dramatically expanded the authority of American law enforcement for the stated purpose of fighting terrorism in the United States and abroad. It has also been used to detect and prosecute other alleged potential crimes, such as providing false information on terrorism. Federal courts declared some sections unconstitutional because they interfere with civil liberties. It was renewed on March 2, 2006 with a vote of 89 to 11 in the Senate and on March 7 280 to 138 in the House. The renewal was signed into law

Richard D. Newbold

by President Bush on March 9, 2006. Some of the more controversial provisions of USA PATRIOT act were largely inspired by the RICO act, which restricted due process for individuals involved in organized crime, racketeering, and drug trafficking.

Payload - Any identification information which is carried inside the biometric data record. It can be used as identification for an entity once biometric authentication is successfully completed.

Penetration rate - The penetration rate is defined as the expected proportion of the template data to be searched over all input samples under the rule that the search proceeds through the entire partition regardless of whether a match is found.

Pentagon - The Pentagon is the headquarters of the United States Department of Defense, located at 48 N. Rotary Road, Arlington, Virginia 22211 (Map). Its mailing address is "Washington, DC 20301." As a symbol of the U.S. military, "the Pentagon" is often used metonymically to refer to the Department of Defense rather than the building itself. Those who work within its walls often simply call it the Building. The building was dedicated on January 15, 1943. It is the highest-capacity office building in the world and one of the world's largest buildings in terms of floor area. It houses approximately 23,000 military and civilian employees and about 3,000 non-defense support personnel. It has five sides, five floors above ground (plus two basement levels), and five ring corridors per floor with a total of 17.5 miles (28 km) of corridors. At five acres (20,000 m²), the central plaza in the Pentagon is the world's largest "no-salute, no-cover" area (an area exempt from the normal rule that, when out of doors, U.S. military personnel must wear hats and salute superior officers). The open space in the center is informally known as "ground zero," a nickname originating during the Cold War when it was thought of as the most likely target of a nuclear missile. At the center of this plaza is the "Ground Zero Cafe," a snack bar. Just south of the Pentagon are Pentagon City and Crystal City, extensive shopping and high-density residential districts in Arlington. Arlington National Cemetery is to the north. The Washington Metro Pentagon station is also located at the Pentagon, on the Blue and Yellow Lines. The Pentagon is surrounded by the complex Pentagon road network, has also been known to locals as "The Mixing Bowl."

Performance - A catch-all phrase for describing a measurement of the characteristics, such as accuracy or speed, of a biometric algorithm or system; indicates the accuracy, speed, and robustness of the system capturing the biometric. *See also accuracy, crossover error rate, cumulative match characteristics, d-prime, detection error tradeoff, equal error rate, false accept rate, false alarm rate, false match rate, false reject rate, identification rate, operational evaluation, receiver operating characteristics, scenario evaluation, technology evaluation, true accept rate, true reject rate, and verification rate.*

Performance criteria - Pre-determined criteria established to evaluate the performance of the biometric system under test.

Permanence - Measures how well a biometric resists aging.

Permissions - Commonly used to refer to the access rights provided by information access control system, in physical access control systems the common term is privileges.

Physical biometric - A biometric, which is characterized by a physical characteristic rather than a behavioral trait.

Physical characteristic - A physically measurable part of the human body.

Physical credential - A document that contains printed identification information and often contains a photograph, signature, or both as evidence of identity and of one's right to credit, confidence, authority or privileges. Examples of physical credentials are the driver license, passport, and security photo ID badges. *See also credential.*

Physical/Physiological biometric - A biometric that is characterized by a physical characteristic rather than a behavioral trait, however behavioral elements may influence the biometric sample captured.

Physical security - That part of security concerned with physical measures designed to safeguard personnel; to prevent or delay unauthorized access to equipment, installations, material and documents; and to safeguard them against espionage, sabotage, damage, and theft.

Physical Security Equipment (PSE) - A generic term encompassing any item, device, or system that is used primarily for the protection of Government property, including nuclear, chemical, and other munitions, personnel, installations, and in the safeguarding of national security information and material, including the destruction of such information and material both by routine means and by emergency destruct methods.

PIV key type - A type of a key, The PIV key types are: 1) PIV authentication key; 2) PIV card authentication key; 3) PIV digital signature key; 4) PIV key management key; and 5) card application administration key.

Pixel - A picture element. This is the smallest element of a display that can be assigned a color value. *See also pixels per inch (PPI) and resolution.*

Pixels Per Inch (PPI) - A measure of the resolution of a digital image. The higher the PPI, the more information is included in the image, and the larger the file size. *See also pixel and resolution.*

Platen - The surface on which a finger is placed during optical finger image capture.

Plug-and-Play - An industry-wide standard for add-on hardware that indicates that it will configure itself, eliminating the need to set jumpers and making installation of the product quick and easy.

Population - The set of potential end users for an application.

Positive claim - A claim by a user to be enrolled in the biometric system, an explicit claim is often accompanied by user identification and may also be associated with a password or PIN.

Positive claim of identity - The claims (either explicitly or implicitly) to be enrolled in or known to the system. An explicit claim might be accompanied by a claimed identity in the form of a name, PIN or identification number. Common access control systems are an example.

Positive identification - The process of finding an identity of a person using only the presented biometric characteristics. No initial claim of identity is needed by this function

Privacy Act - Unofficial name for THE PRIVACY ACT OF 1974 5 U.S.C. § 552a.

Privacy-Invasive - A privacy-invasive system facilitates or enables the usage of personal data in a fashion inconsistent with generally accepted privacy principles.

Privacy-Neutral - A privacy-neutral system is one in which privacy is not an issue, or in which the potential privacy impact is slight. Privacy-neutral systems are difficult to misuse from a privacy perspective, but do not have the capability to protect personal privacy.

Privacy-Protective - A privacy-protective system is one used to protect or limit access to personal information, or which provide a means for an individual to establish a trusted identity.

Privacy-Sympathetic - A privacy-sympathetic system is one that limits access to and usage of personal data and in which decisions regarding design issues such as storage and transmission of biometric data are informed, if not driven, by privacy concerns.

Private Key - The part of a key pair to be safeguarded by the owner. A private key is used to generate a digital signature. Private keys are used to decrypt information, including key encryption keys during key exchange. It is computationally infeasible to determine a private key given the associated public key.

Privileges - Term commonly used to refer to the access rights provided by physical access control systems. In information access control systems, the common term is permissions.

Probe - The biometric sample that is submitted to the biometric system to compare against one or more references in the gallery. *See also gallery.*

Protection context - Contains all the information needed to apply cryptographic enhancements to a biometric descriptor, or to remove any existing protection from a biometric descriptor.

Protection profile - 1) Common criteria specification that represents an implementation-independent set of security requirements for a category of Target of Evaluations that meets specific consumer needs. 2) A form of generic Security Target defined in the Common Criteria. Large scale identification - (i.e., "Is this person in the database?"). Given an input biometric sample, a large-scale identification determines if the pattern is associated with any of a large number (e.g., millions) of enrolled identities. Typical large-scale identification applications include welfare-disbursement, national ID cards, border control, voter ID cards, driver's license, criminal investigation, corpse identification, parenthood determination, missing children identification, etc. These large-scale identification applications require a large sustainable throughput with as little human supervision as possible.

Provisioning - To provide users (such as the cardholders in an access control system or the users of a computer-based information system) with two things: 1) a means to authenticate themselves (such as a card and PIN, or name and password), and 2) access privileges. Those two elements combined (a means to authenticate and privileges) are what enable access to protected assets. *See also user provisioning.*

Public key - 1) The Part of a key pair that is made public, usually by posting it to a directory. A public key can be either a signature or key exchange key. The signer's public signature key is used to verify a digital signature. Sending an encrypted message requires use of the recipient's public key in the encryption process. 2) The published key of a public/private key pair.

Public key cryptography - Public key cryptography is a form of cryptography which generally allows individuals or systems to communicate securely without having prior access to a shared secret key (symmetric key). This is done by using a pair of cryptographic keys, designated as public key and private key, which are related to each other mathematically. What one encodes with one key one can decode only with the other key, and vice-versa. Yet one cannot discern one key if one has the other key. This allows one key to be made public without risking disclosure of the other key, which is kept private. Thus the two cryptographic keys are known as a "public key/ private key pair." Public/private key pairs have a number of uses, including encryption and the computations involved in creating and verifying digital signatures. Public key cryptography is also known as asymmetric cryptography, because a different key is used to decode the information than was used to encode it. Private and public keys are often referred to as asymmetric private keys, asymmetric public keys, or simply asymmetric keys to refer to them both.

Public key digital signature - *See digital signature.*

Public key encryption - Encryption using a public/private key pair. *See also public key cryptography.*

Public key infrastructure (PKI) - PKI is a security management system including hardware, software, people, processes and policies (including certificate authorities and registration authorities) dedicated to the management of digital certificates for the purpose of achieving secure exchange of electronic information. The term PKI is also sometimes used loosely simply as a reference to public key cryptography. Because a digital certificate contains the public key of the subscriber (the person the certificate was issued to), it is sometimes also called a public key certificate or PKI certificate (FIPS 201 uses all three terms).

Public versus Private - Question of whether the users of the system be customers of the system management (public) or employees (private). Attitudes toward usage of the devices, which will directly affect performance, vary depending upon the relationship between the end-users and system management.

Q

Q - Number of matches required by decision policy to declare an identification.

QA - Quality Assurance

QC - Quality Control

QNSP - Questionnaire for National Security Positions

QT-CBA - Quick Turn Capabilities Based Assessment

R

R&D - Research and Development

RA - Registration Authority

RAPIDS - Real-time Automated Personnel Identification System

RBAC - Role Based Access Control - The basic concept of RBAC is that within an organization, roles are created for various job functions, and personnel are assigned a specific role. Corresponding roles are created in the access control system, and access privileges are assigned to the roles (as opposed to being assigned directly to personnel). Thus personnel acquire access privileges by being assigned a role. This use of roles facilitates policy-based management of access control that mirrors the actual job requirements of an organization's personnel.

RD - Restricted Data

RDECOM - (Army) Research, Development, Engineering Command

RDT&E - Research, Development, Test, and Evaluation

REL - This Information Has Been Authorized for Release to (UK, CAN, etc.)

RF (capture) - A unique type of finger image capture that uses RF signals to capture the finger image under the outer layer of the skin, to the live layer below.

RFA - Report for Adjudication

RFC - Request for Comment

RFI - Representative of a Foreign Interest

RFID - Radio Frequency Identification - 1) RFID and other wireless technologies involve the transmission of information through the open air. When these technologies are used to transfer PII or are associated in any way with individuals, these technologies raise privacy issues regarding surveillance and involuntary identification. The broadcast nature of the transmission and the association of that data traffic with an individual raises privacy concerns that should be addressed early in the project life cycle. An RFID system typically includes three elements: a tag, a reader, and a database. An RFID tag or transponder comprises a chip that contains a unique number that identifies an object (and perhaps other information) and is connected to an antenna. Each antenna enables the chip to communicate via radio waves to a reader, which captures the unique number or other data on the tag. That data can then be transmitted to computers that store information about the object to which the tags are attached. In most instances, the protocol for communication between the reader and tag enables a fixed set of commands. Tags typically do not have the capacity to upload and execute additional software programs. 2) Technology that uses low-powered radio transmitters to read data stored in a transponder (tag). RFID tags can be used to track assets, manage inventory, authorize payments, and serve as electronic keys. RFID is not a biometric.

RFP - Request for Proposal

RFU - Reserved for Future Use

RID - Registered application provider Identifier

RISC - Repository for Individuals of Special Concern

ROC - Receiver Operating Characteristic

ROC (also) - Receiver Operating Curves - 1) A method of showing measured accuracy performance of a biometric system. 2) A graph showing how the false rejection rate and false acceptance rate vary according to the threshold.

ROI - Report of Investigation

ROMO - Range of Military Operations

RON - Report of National Agency Check

RP - Relying Party

Raw - Image file format called bit-map in which the image is stored in the same format in which it is stored in video memory, typically one byte (for monochrome images) per picture element or three bytes (for color images) per picture element.

Raw biometric data - The captured, unprocessed biometric data (e.g., fingerprint image or audio stream) from a sensor device, in digital form, suitable for subsequent processing to create a biometric sample or template.

Recognition - A generic term used in the description of biometric systems (e.g., face recognition or iris recognition) relating to their fundamental function. The term "recognition" does not inherently imply verification, closed-set identification, or open-set identification (watch list). *See also identification.*

Recognition (of identity) - The operation of comparing a submitted biometric sample against the population of biometric reference templates to determine whether it belongs to the population and which member of the population it is.

Recognition attempt - A recognition attempt is the presentation of a single biometric sample to a biometric identification device for an identification decision.

Record - 1) Under the Privacy Act, any item, collection, or grouping of information about an individual that is maintained by an agency, including, but not limited to, his education, financial transactions, medical history, and criminal or employment history and that contains his name, or the identifying number, symbol, or other identifying particular assigned to the individual, such as a finger or voice print or a photograph. 2) The template and other information about the end user (e.g., name, access permissions).

Red Force database - A database containing biometrics and information for known unfriendly forces (e.g., terrorists, criminals, etc.)

Re-enrollment - The process of enrolling an individual's biometric data where the same or other biometric data has been enrolled at least once. *See also enrollment and initial enrollment.*

Reference - The biometric data stored for an individual for use in future recognition. A reference can be one or more templates, models or raw images. *See also template.*

Reference template - Processed biometric data stored as representative of the user's biometric sample.

Registration - The process of making a person's identity known to a biometric system, associating a unique identifier with that identity, and collecting and recording the person's relevant attributes into the system.

Registration Authority (RA) - The person or company responsible for the identification and authentication of digital certificate subscribers prior to certificates being issued by the certification authority. The registration authority does not sign or issue the certificates (the certificate authority does). The registration authority is responsible for the accuracy of the information contained in a certificate request.

Relying party - An entity that relies upon the subscriber's credentials, typically to process a transaction or grant access to information or a system.

Resolution - The number of pixels per unit distance in the image. Describes the sharpness and clarity of an image. *See also pixel and pixels per inch (PPI).*

Response time - The time period for a biometric system to return a decision on identification or verification of a biometric sample; the time that is required to complete a biometric transaction with a given device.

Retina - A physical biometric that analyses the layer of blood vessels situated at the back of the eye.

Ridge ending - 1) Occurs at the point on a fingerprint or palm print where a friction ridge begins or ends without splitting into two or more continuing ridges. 2) A minutiae point at the ending of a friction ridge. *See also bifurcation and friction ridge.*

Robustness - A characterization of the strength of a security function, mechanism, service, or solution, and the assurance (or confidence) that it is implemented and functioning correctly.

Rolled fingerprints - An image that includes fingerprint data from nail to nail, obtained by "rolling" the finger across a sensor.

S

(S) - Secret

S - System throughput rate

S&T - Science & Technology

SaaS - Soldier as a System

SAC - Senate Appropriations Committee

SAE - Service Acquisition Executive

SAEWG - Security Awareness & Education Working Group

SAFE - Signatures and Authentication for Everyone

SAFTI - Secure and Facilitated International Travel Initiative

SAIC - Science Applications International Corporation

SAML - Security Assertion Mark-up Language

SAO - Special Access Office

SAP - Special Access Program

SAP (also) - Systems Acquisition Plan

SAPCO - Special Access Program Coordinating Office

SAPF - Special Access Program Facility

SAPOC - Special Access Program Oversight Committee

SAPWG - Special Access Program Working Group

SAR - Special Access Required

SASC - Senate Armed Services Committee

SAV - Staff Assistance Visit

SAV (also) - Security Assistance Visit

SBI - Secure Border Initiative

SBI (also) - Special Background Investigation - An investigation conducted by DSS, with extended coverage of the individual's background to provide a greater depth of knowledge than a BI. The scope of an SBI is 15 years or since the 18th birthday, whichever is shorter. At least two years will be covered, except that no investigation is conducted prior to the subject's 16th birthday.

SBIR - Small Business Innovation Research - The purpose of the DoD's Small Business Innovation Research program is to harness the innovative talents of our nation's small technology companies for U.S. military and economic strength.

SBU - Sensitive But Unclassified

SCEPACS - Smart Card Enabled Physical Access Control System

SCG - Security Classification Guide

SCI - Sensitive Compartmented Information - SCI information is usually only briefed, discussed, and stored in an accredited SCI facility. Moreover, programs handled under the SCI paradigm are normally not publicly acknowledged by the U.S. government.

SCIF - Sensitive Compartmented Information Facility - An enclosed area within a building that is used to process Sensitive Compartmented Information level classified information. A SCIF can also be located in a mobile configuration and can be deployed using air, ground or maritime resources. The physical construction, access control, and alarming of the facility is defined under Director of Central Intelligence Directive (DCID) 6/9, and was previously specified through DCID 1/21. The computers running within this facility must operate under rules set forth in DCID 6/3. Computers and telecommunication equipment within must fall within the TEMPEST emanations specification.

SCM - Security Countermeasures

SCVP - Server-based Certificate Validation Protocol

SDI - Strategic Defense Initiative

SDIO - Strategic Defense Initiative Organization

SDK - Software Developer's Kit - A programming package that enables a programmer to develop applications for a specific platform, typically an SDK includes one or more APIs, programming tools, and documentation.

SDO - Standards Development Organization

SDR - System Design Review

SECDEF - Secretary of Defense

SECNAVINST - Secretary of the Navy Instruction

SEIWG-012 - Federal standard for security card identification that defines a numerical sequence of 40 digits containing several different numbers such as an "agency code" and a "credential code". It is named after the group that developed it, the Security Equipment Integration Working Group (SEIWG), a sub-group of the Physical Security Equipment Action Group (PSEAG), which is a DoD organization that coordinates all of the physical security research and development efforts across the armed services. FIPS 201 specifies a new standard that replaces the SEIWG-012, the Federal Agency Smart Credential Number (FASC-N).

SERC - Software Engineering Research Center

SES - Senior Executive Service

SEVIS - Student and Exchange Visitor Information System

SF - Special Forces

SF (also) - Security Forces

SF 311 - Agency Information Security Program Data

SF 312 - Classified Information Nondisclosure Agreement

SF 328 - Certificate Pertaining to Foreign Interests

SF 700 - Security Container Information

SF 701 - Activity Security Checklist

SF 702 - Security Container Check Sheet

SF 703 - TOP SECRET Cover Sheet

SF 704 - SECRET Cover Sheet

SF 705 - CONFIDENTIAL Cover Sheet

SF 706 - TOP SECRET (Label)

SF 707 - SECRET (Label)

SF 708 - CONFIDENTIAL (Label)

SF 709 - CLASSIFIED (Label)

SF 710 - UNCLASSIFIED (Label)

SF 711 - DATA DESCRIPTOR (Label)

SF 85 - Questionnaire for Non-sensitive Positions

SF 85P - Questionnaire for Public Trust Positions

SF 85P-S - Supplemental Questionnaire for Selected Positions

SF 86 - Questionnaire for National Security Positions

SF 87 - Fingerprint Card

SG - Study Group

SI - Special Intelligence

SIB - State Information Bureau

SIF - Special/Suitability Issue File

SIGINT - Signals Intelligence

SIGSEC - Signals Security

SII - Special Investigative Inquiry

SII (also) - Security/Suitability Investigations Index (OPM)

SII (also) - Statement of Intelligence Interest

SIO - Senior Intelligence Officer

SIOP - Single Integrated Operational Plan

SIOP/ESI - Single Integrated Operational Plan/Extremely Sensitive Information

SIPRNet - Secret (formerly Secure) Internet Protocol Router Network - A system of interconnected computer networks used by the U.S. Department of Defense and the U.S. Department of State to transmit classified information (up to and including information classified SECRET) by packet switching over the TCP/IP protocols in a "completely secure" environment. It also provides services such as hypertext documents and electronic mail. In other words, the SIPRNet is the DoD's classified version of the civilian Internet together with its counterpart, the Top Secret and SCI Joint Worldwide Intelligence Communications System, JWICS.

SISSU/C&A - Security, Interoperability, Supportability, Sustainability, Usability/Certification and Accreditation

SLA - Service Life Allowance

SME - Subject Matter Expert

SMT - Scars, Marks, & Tattoos

SMTP - Simple Mail Transfer Protocol

SOCOM - U.S. Special Operations Command

SOIC - Senior Official of the Intelligence Community

SON - Statement of Need

SOP - Standard Operating Procedure/Plan

SOR - Sexual Offender Registry

SOR (also) - Statement of Reasons

SORN - System of Records Notice

SOUTHCOM - U.S. Southern Command

SOW - Statement of Work

SP - Security Police

SP (also) - Service Provider

SP (also) - Special Publication

SPB - Security Policy Board

SPF - Security Policy Forum

SPG - Security Procedures Guide

SPO - Systems Program office

SPOC - Special Access Program Oversight Committee (USAF)

SPOT - Synchronized Predeployment and Operational Tracker

SPP - Standard Practice Procedures

SPP (also) - Security Procedures Plan

S-PR - Secret Periodic Reinvestigation

SPRG - (Navy) Special Access Program Review Group

SPT - Simplified Passenger Travel

SRG - Senior Review Group

SRR - System Requirements Review

SSBI - Single Scope Background Investigation

SSCI - Senate Select Committee on Intelligence

SSDC - Space and Strategic Defense Command

SSL - Secure Socket Layer - A protocol that allows secure communications on the World Wide Web/Internet.

SSM - System Security Manager

SSMP - System Security Management Plan

SSO - Special Security Officer

SSO (also) - Single Sign-On

SSP - Shared Service Provider

ST - Security Target - A set of security requirements and specifications to be used as the basis for evaluation of an identified Target of Evaluation.

ST&E - Security Test and Evaluation

STAR - System Threat Assessment Report

START - Strategic Arms Reduction Treaty

STE - Secure Terminal Equipment

STINFO - Scientific and Technical Information

STOPSO - Strategic Operational Solutions, Inc.

STR - Short Tandem Repeats - STR analysis uses nuclear DNA that is inherited from both parents and is inherited from both parents and is stored in cells in the form of 23 pairs of chromosomes. STRs are 2-6 base pair sequence repeats that vary in number across individuals.

STRATCOM - U.S. Strategic Command

STU - Secure Telephone Unit

SW1 - First byte of a two-byte status word.

SW2 - Second byte of a two-byte status word.

SYSCOM - Systems Command

Sample - A biometric measure presented by the user and captured by the data collection subsystem as an image or signal (e.g., fingerprint, face, and iris images).

Scenario Evaluation - One of the three types of performance evaluations. The primary goal of a scenario evaluation is to measure performance of a biometric system operating in a specific application. *See also technology evaluation and operational evaluation.*

Scolara (or Sclera) - Consists of closely interwoven fibers, and covers over the entire surface of the eye, except for a small section in the back, where the optic nerve leaves the eye, and a small section directly front and center, known as the cornea. The sclera is currently being studied by CITeR as a biometric modality that may lend to improved iris recognition by taking advantage of the already available conjunctival vasculature patterns of the sclera.

Score - 1) A number indicating the degree of similarity or correlation of a biometric match. Traditional authentication methods—passwords, PINs, keys, and tokens—are binary, offering only a strict yes/no response. This is not the case with most biometric systems. Nearly all biometric systems are based on matching algorithms that generate a score subsequent to a match attempt. This score represents the degree of correlation between the verification template and the enrollment template. There is no standard scale used for biometric scoring: for some vendors a scale of 1-100 might be used, others might use a scale of 1 to 1. Some vendors may use a logarithmic scale and others a linear scale. Regardless of the scale employed, this verification score is compared to the system's threshold to determine how successful a verification attempt has been. 2) The level of similarity from comparing a biometric sample against a previously stored template. 3) A value indicating the degree of similarity or correlation between a biometric sample and a reference template.

SENTRI/DCL - Secure Electronic Network for Travelers Rapid Inspection/Dedicated

Segmentation - The process of parsing the biometric signal of interest from the entire acquired data system. For example, finding individual finger images from a slap impression.

Sensor - The physical hardware device used for biometric capture.

Sensor - Hardware found on a biometric device that converts biometric input into a digital signal and conveys this information to the processing device; Hardware found on a biometric device that converts biometric input into electrical signals and conveys this information with the attached computer (e.g., fingerprint sensor).

Sensor ageing - The gradual degradation in performance of a sensor over time.

Signature dynamics - A behavioral biometric modality that analyzes dynamic characteristics of an individual's signature, such as shape of signature, speed of signing, pen pressure when signing, and pen-in-air movements, for recognition.

Signature verification - A behavioral biometric that analyses the way an end user signs his/her name. The signing features such as speed, velocity and pressure exerted by a hand holding a pen are as important as the static shape of the finished signature.

Similarity score - A value returned by a biometric algorithm that indicates the degree of similarity or correlation between a biometric sample and a reference. *See also difference score and hamming distance.*

Single error rates - Error rates state the likelihood of an error (false match, false non-match, or failure to enroll) for a single comparison of two biometric templates or for a single enrollment attempt. This can be thought of as a single error rate.

Single factor authentication - Authentication using only one identity factor of the following three: 1) Knowledge factor - something an individual knows; 2) Possession factor - something an individual has; or 3) Biometric factor - something an individual is.

Skimming - The act of obtaining data from an unknowing end user who is not willingly submitting the sample at that time. An example could be secretly reading data while in close proximity to a user on a bus. *See also eavesdropping.*

Slap fingerprint - Fingerprints taken by simultaneously pressing the four fingers of one hand onto a scanner or a fingerprint card. Slaps are known as four finger simultaneous plain impressions.

Software type - An implementation type in a matching algorithm.

Speaker recognition evaluations - An ongoing series of evaluations of speaker recognition systems.

Speaker separation - A technology that separates overlapping voices from each other and other background noises.

Speaker-dependent - A term sometimes used by speaker verification vendors to emphasize the fact their technology is designed to distinguish among voices.

Speech recognition - 1) This is not generally considered a biometric and should not be confused with speaker verification. Speech recognition involves recognizing words as they are spoken and does not identify the speaker. 2) A biometric modality that uses an individual's speech, a feature influenced by both the physical structure of an individual's vocal tract and the behavioral characteristics of the individual, for recognition purposes. Sometimes referred to as "voice recognition." Speech recognition recognizes the words being said. 3) A technology that enables a machine to recognize spoken words. *See also speaker recognition and voice recognition.*

Speaker verification - Speaker verification is performed by computing principal components of a fixed text statement comprising a speaker identification code and a two-word phrase, and principal spectral components of a random word phrase. A multi-phrase strategy is utilized in access control to allow successive verification attempts in a single session, if the speaker fails initial attempts. Based upon a verification attempt, the system produces a verification score which is compared with a threshold value. On successive attempts, the criterion for acceptance is changed, and one of a number of criteria must be satisfied for acceptance in subsequent attempts.

A speaker normalization function can also be invoked to modify the verification score of persons enrolled with the system who inherently produce scores which result in denial of access. Accuracy of the verification system is enhanced by updating the reference template which then more accurately symbolizes the person's speech signature.

Spoofing - The ability to fool a biometric sensor into recognizing an illegitimate user as a legitimate user (verification) or into missing an identification of someone that is in the database. *See also liveness detection and mimic.*

Standard environment - A fifth partition is "standard/non-standard operating environment." If the application will take place indoors at standard temperature (20° C), pressure (1 atm.), and other environmental conditions, particularly where lighting conditions can be controlled, it is considered a "standard environment" application. Outdoor systems, and perhaps some unusual indoor systems, are considered "non-standard environment" applications.

Standoff - Biometric identification is sometimes better made at a distance, especially if the impostor poses a physical threat. This can be done in a cooperative setting where the scanner/reader could be at a "safe" distance.

State - Refers to the Department of State or U.S. State Department.

Static biometric verification method - A biometric verification method which requires the presentation of a physiological (i.e., static) feature of a person to be authenticated (see "type A") or performance of an enrolled, pre-determined action.

Static signature verification - Verification of signature based only on the shape of the resulting signature.

Status word - Two bytes returned by an integrated circuit card after processing any command that signify the success of or errors encountered during said processing.

Submission - The process whereby an end user provides a biometric sample to a biometric system; the process whereby a user provides behavioral or physiological data in the form of biometric samples to a biometric system. A submission may require looking in the direction of a camera or placing a finger on a platen. Depending on the biometric system, a user may have to remove eyeglasses, remain still for a number of seconds, or recite a pass phrase in order to provide a biometric sample. *See also capture.*

Subscriber - *See digital certificate subscriber.*

Supplementary data - A generic term referring to data that is used supplementarily during enrollment or verification. Normalized data is a piece of supplementary data.

Supplementary data set - A collection of one or more piece of supplementary data.

Supplementary signature data - Signature data to be used for generation of supplementary data that are, in turn, used supplementary during enrollment or verification.

Supplementary voice data - Voice data to be used for generation of supplementary data that are, in turn, used supplement during enrollment or verification.

Symmetric cryptography - Two parts of a thing are similar. In symmetric cryptography the same key is used for both encrypting and decrypting data.

Symmetric key - Encryption methodology in which the encryptor and decryptor use the same key, which must be kept secret.

Synchronous multimodality - The use of multiple biometric technologies in a single authentication process. For example, biometric systems exist which use face and voice simultaneously, reducing the likelihood of fraud and reducing the time needed to verify.

System identifier(s) - Identifiers used for system-to-system electronic communications across the enterprise. They are not to be declared by, nor in fact generally known to, the person to whom they are assigned. Their primary purpose is to limit the ambiguity in identity caused by human entry of declarative identifiers (e.g., transpositions and typographical errors that occur when entering SSNs). Once they are assigned they are used only for technology-to-technology communications and never printed on any media. Their scope is only for use within the Department of Defense.

Synthetic identity fraud - A type of ID fraud in which thieves literally create new identities either by combining real and fake identifying information to establish new accounts with fictional identities or create the new identity from totally fake information. In typical synthetic fraud, a fraudster uses a real Social Security number and combines it with a name other than the one associated with that number. The combination often doesn't hit the consumer's credit report.

System of records - A group of any records under the control of any agency from which information is retrieved by the name of the individual or by some identifying number, symbol, or other identifying particular assigned to the individual.

Sytem of Records Notice (SORN) - Published in the Federal Register before a system of records goes into operation.

T

T - Number of independent templates or models stored as an ensemble for each enrolled user

T&E - Test and Evaluation

TA/CP - Technical Assessment/Control Plan

TAS - Threat Analysis Section

TASO - Terminal Area Security Officer

TBCMS-M - Tactical Biometrics Collection and Matching System-Maritime

TBS - Theater Badging System

TC - Technical Committee

TCN - Third Country Nation - Individuals who are not of Coalition or Iraq nationality as determined by their passport of identification card.

TCO - Termination Contracting Officer

TCP - Technology Control Plan

TCP/IP - Transmission Control Protocol/Internet Protocol - A protocol for communication between computers, used as a standard for transmitting data over networks and the Internet.

TEDAC - Terrorist Explosives Device Analytical Center - TEDAC is an FBI-led initiative, created in December 2003 committed to establishing that global response by creating a single federal program responsible for the worldwide collection, complete forensic and technical analysis and timely dissemination of intelligence regarding terrorist bombs. Information gleaned from TEDAC's analysis is shared throughout the law enforcement, intelligence and military communities.

TEMP - Test and Evaluation Master Plan

TESOP - Test and Evaluation Standard Operating Procedure

TIARA - Tactical Intelligence and Related Activity

TIDE - Terrorist Identities Datamart Environment - Serves as the central all-source information on international terrorist identities for use by the U.S. TIDE distributes a "sensitive but unclassified" extract to the Terrorist Screening TSC, in turn, validates this information and provides it to Federal departments foreign governments that use this information to screen for terrorists.

TIG - Technical Implementation Guidance

TIP - Terrorist Interdiction Program - Enhances border security capabilities of those countries at risk for terrorist activity. TIP provides participating countries with a computerized watchlisting system to help constrain terrorist mobility globally.

TIW - Transnational Infrastructure Warfare - TIW is the attacking of a nation's key industries and utilities, including telecommunications, banking and finance, transportation, water, government operations, emergency services, energy and power, and manufacturing.

TLV - Tag-Length-Value

TMA - TRICARE Management Agency

TMC- Two-Man Control

TMO - (Army) Technology Management Office

TnC - Transnational Criminal

TOE - Target of Evaluation - An information technology product or system and its associated administrator and user guidance documentation that is the subject of an evaluation.

TOR - Terms of Reference

TOST - Tactical Overwatch Support Team

TOT - Type-of-Transaction

TPDC - Training and Personnel Development Committee

TPI - Two-Person Integrity

TPS - Transportation Protection Service

TRADOC - Training and Doctrine Command

TRANSCOM - U.S. Transportation Command

TRDP - Technical Research and Development Program

TRL - Technology Readiness Level

(TS) - Top Secret

TSA - The Transportation Security Administration - A controversial U.S. government agency that was created as part of the Aviation and Transportation Security Act passed by the U.S. Congress and signed into law by President George W. Bush on November 19, 2001. The TSA was originally organized in the U.S. Department of Transportation but was moved to the U.S. Department of Homeland Security on March 1, 2003.

TSAG - Telecommunication Standardization Advisory Group

TSB - Telecommunication Standardization Bureau

TSCM - Technical Surveillance Countermeasures

TSCO - Top Secret Control Officer

TSEC - Telecommunications Security

TSWG - Technical Support Working Group - Co-chaired by DOD and State, TSWG is the national forum to identify, prioritize, and coordinate U.S. Government Interagency and international research and development requirements to combat terrorism.

TT - Technology Transfer

TTE - Tactics, Techniques, and Procedures

TTI - Technology Transition Initiative - Congress established the Technology Transition Initiative in 2003 to help bridge the funding gap between demonstration and production of DoD Science and Technology (S&T) funded technology.

TWIC - Transportation Worker Identification Credential - TWIC enrollment began at the Port of Wilmington, Delaware October 16, 2007. TWIC's purpose is to ensure individuals who pose a threat do not gain unescorted access to secure areas of the nation's maritime transportation system. TWIC was established by Congress through the Maritime Transportation Security

Act (MTSA) and is administered by the Transportation Security Administration (TSA) and U.S. Coast Guard. TWICs are tamper-resistant biometric credentials for workers who require unescorted access to secure areas of ports, vessels, outer continental shelf facilities and all credentialed merchant mariners. It is anticipated that more than 750,000 workers including longshoremen, truckers, port employees and others will be required to obtain a TWIC. To obtain a TWIC, an individual must provide biographic and biometric information such as fingerprints, sit for a digital photograph and successfully pass a security threat assessment conducted by TSA.

Technical data - Recorded information (regardless of the form or method of the recording) of a scientific or technical nature (including computer software documentation) relating to supplies procured by an agency. Such term does not include computer software or financial, administrative, cost or pricing, or management data or other information incidental to contract administration.

Technology evaluation - One of the three types of performance evaluations. The primary goal of a technology evaluation is to measure performance of biometric systems, typically only the recognition algorithm component, in general tasks. *See also operational evaluation, scenario evaluation.*

Template - A digital representation of an individual's distinct characteristics, representing information extracted from a biometric sample. Templates are used during biometric authentication as the basis for comparison. *See also extraction, feature, and model.*

Template (also) - 1) Data that represents the biometric measurement of an enrollee, used by a biometric system for comparison against subsequently submitted biometric samples. 2) A mathematical representation of biometric data. A template can vary in size from nine bytes for hand geometry to several thousand bytes for facial recognition. 3) A biometric data structure comprised of processed enrollment data, suitable for storage and subsequent matching. 4) Data, which represents the biometric measurement of an enrollee, used by a biometric system for comparison against subsequently submitted biometric samples. 5) A user's stored reference measure based on biometric feature(s) extracted from biometric sample(s). 6) Value field of a constructed data object, defined to give a logical grouping of data objects, as defined by ISO/IEC 7816-6.

Template/Reference template - Data which represents the biometric measurement of an enrollee, used by a biometric system for comparison against subsequently submitted biometric samples. The biometric template can either be a normalized biometric image or an approximation.

Template ageing - The degree to which biometric data evolves and changes over time, and the process by which templates account for this change; the gradual change of a user's biometric feature(s) which requires periodic updating of the user's reference template.

Template dormant time - The elapsed time between the creation, or last update, of a template and its current use.

Template identifier - The item used to tie the to their own to make sure the correct template is called up when required; this can be a PIN, password, or card number.

Template matching - Where the newly captured sample goes through the same algorithms as the sample captured during the enrolment process.

Template maturity - The number of biometric samples, including creation contributing to the biometric reference template.

Template size - The amount of computer memory taken up by the biometric data.

Template store - The logical component of a biometric system responsible for the storage of templates.

Template updating - The process by which the template data are updated to compensate for a changing template.

Text dependent system - A system that requires a speaker to say a specific set of numbers or words.

Text independent system - A system that creates voiceprints from unconstrained speech and does not require a speaker to say a specific set of numbers or words.

Text prompted system - A speaker verification system that prompts the speaker to say randomly ordered numbers or words. The term "challenge-response" is also used in a similar way to define text prompting.

Thermal - A finger image capture technique that uses a sensor to sense heat from the finger and thus capture a finger image pattern.

Third party test - An objective test, independent of a biometric vendor, usually carried out entirely within a test laboratory in controlled environmental conditions.

Threat - An intentional or unintentional potential event that could compromise the security and integrity of the system. *See also vulnerability.*

Threshold - 1) A user setting for biometric systems operating in the verification or open-set identification (watch list) tasks. The acceptance or rejection of biometric data is dependent on the match score falling above or below the threshold. The threshold is adjustable so that the biometric system can be more or less strict, depending on the requirements of any given biometric application; a predefined number, often controlled by a biometric system administrator, which establishes the degree of correlation necessary for a comparison to be deemed a match. 2) A parametric value used to convert a matching score to a decision; the value by which it is decided whether verification data matches enrollment data with reference to their matching score. 3) A value to discriminate whether the enrollment data and verification data match relative to the matching score or distance. 4) A predefined value which establishes the degree of similarity or correlation (i.e., score) necessary for a biometric sample to be deemed a match with a reference template. The acceptance or rejection of biometric data is dependent on the match score falling above or below the threshold. The threshold is adjustable so that the biometric system can be more or less strict, depending on the requirements of any given biometric. A threshold change will usually change both FAR and FRR. As FAR decreases, FRR increases. *See also comparison, match, and matching.*

Throughput - The number of biometric transactions that a biometric identification device completes within a stated time interval as evaluated by the applicable IBA test procedure specified.

Throughput rate - The number of biometric transactions that a biometric system processes within a stated time interval.

Token - 1) A possession that shows the identity of its owner such as a smart card, a 2D bar-code on a physical support or a contactless card. 2) A physical device that contains information specific to the user; a physical object that indicates the identity of its owner, for example, a smart card.

Transaction - An attempt by a user to validate a claim of identity or non-identity by consecutively submitting one or more samples, as allowed by the system decision policy.

Treasury - Refers to the Treasury Department or U.S. Department of the Treasury

Triradius - A point located at the meeting point of three directional ridges. Its count varies from pattern to pattern, none for simple arch, one for tented arch, loop and two or more for whorl. In the palm of the hand triradii are marked by letters like *a, b, c, d or t*.

True accept rate - A statistic used to measure biometric performance when operating in the verification task. The percentage of times a system (correctly) verifies a true claim of identity. For example, Frank claims to be Frank and the system verifies the claim.

True-name identity fraud - Consumer's real identifying information is used without modification. The fraudster poses as the actual consumer. *See also synthetic identity fraud.*

True reject rate - A statistic used to measure biometric performance when operating in the verification task. The percentage of times a system (correctly) rejects a false claim of identity. For example, Frank claims to be John and the system rejects the claim.

Type I error - An error that occurs in a statistical test when a true claim is (incorrectly) rejected. For example, John claims to be John, but the system incorrectly denies the claim. *See also false reject rate (FRR).*

Type II error - An error that occurs in a statistical test when a false claim is (incorrectly) not rejected. For example, Frank claims to be John and the system verifies the claim. *See also false accept rate (FAR).*

U

(U) - Unclassified

U - Number of active, enrolled users

U.S.C. - United States Code

U-actors - Unintentional cyber actors - Individuals who unintentionally attack but affect national security and are largely unaware of the international ramifications of their actions. Unintentional actors may be influenced by I-actors but are unaware they are being manipulated to participate in cyber operations. U-actors include anyone who commits CyI, CyM, CyA, and CyR without the intent to affect national security or to further operations against national security. This group also includes individuals involved in CyC, journalists, and industrial spies. The threat of journalists and industrial spies against systems including unintentional attacks caused by their CyI efforts should be considered high.

UA - Unintentional cyber warfare attack - Unintentional attack is basically crime. UA may be committed by a novice hacker or a professional cyber criminal, but the intent is self-serving and not to further any specific national objective. This does not mean unintentional attacks cannot affect policy, or have devastating effects.

UGS - Unattended Ground Sensor - UGS consist of a variety of sensor technologies that are packaged for deployment and perform the mission of remote target detection, location and/or recognition. Ideally, the UGS are small, low cost and robust, and are expected to last in the field for extended periods of time after deployment. They are capable of transmitting target information back to a remote operator. These devices could be used to perform various mission tasks including perimeter defense, border patrol and surveillance, target acquisition, and situation awareness.

UICC - Universal Integrated Circuit Card

ULF - Unsolved Latent File

URI - Uniform Resource Identifier

URL - Uniform Resource Locator

USA - U.S. Army

USACIC - U.S. Army Criminal Investigation Command

USAF - U.S. Air Force

USAR - U.S. Army Reserves

USB - Universal Serial Bus - An interface incorporating a high-speed external bus for personal computers. A USB token is a device containing an embedded microprocessor IC that interfaces directly with a personal computer's USB port without additional hardware (e.g., card reader).

USCG - United States Coast Guard

USCIS - U.S. Citizenship and Immigration Services

USD(A&T) - Under Secretary of Defense (Acquisition and Technology)

USD(I) - Under Secretary of Defense for Intelligence

USD(P) - Under Secretary of Defense (Policy) - Through the Deputy Under Secretary of Defense for Policy (DUSD(P)), develops overall physical security policy, evaluates and validates PSE requirements in relation to policy decisions and recommend to the DUSD(TWP) changes, as necessary, provides the DoD member on the Interagency Advisor Committee on Security Equipment (IACSE), General Services Administration (GSA), Co-chairs the PSESG, provides representation on the PSEAG, and coordinates information security equipment development efforts by the DoD Security Institute with the PSEAG.

USD(P&R) - Under Secretary of Defense for Personnel and Readiness

USEUCOM - United States European Command

USJFCOM - United States Joint Forces Command

USMC - United States Marine Corps

USN - United States Navy

USNORTHCOM - United States Northern Command

USPACOM - United States Pacific Command

USSOCOM - United States Special Operations Command

USSOUTHCOM - United States Southern Command

US-VISIT - U.S. Visitor and Immigrant Status Indicator Technology **-** A continuum of security measures that begins overseas, at the Department of State's visa issuing posts, and continues through arrival and departure from the United States of America. Using biometric, such as digital, inkless finger scans and digital photographs, the identity of visitors requiring a visa is now matched at each step to ensure that the person crossing the U.S. border is the same person who received the visa. For visa-waiver travelers, the capture of biometrics first occurs at the port of entry to the U.S. By checking the biometrics of a traveler against its databases, US-VISIT verifies whether the traveler has previously been determined inadmissible, is a know security risk (including having outstanding wants and warrants), or has previously overstayed the terms of a visa. These entry and exit procedures address the U.S. critical need for tighter security and ongoing commitment to facilitate travel for the millions of legitimate visitors welcomed each year to conduct business, learn, see family, or tour the country.

Ultrasound - A technique for finger image capture that uses acoustic waves to measure the density of a finger image pattern.

Unavailability - 1) A generic term expressing both enrollment unavailability and verification unavailability. 2) A case in which a certain iris or individual cannot be used for performance evaluation due to failure to generate a piece of enrollment data or verification data. The determination as to whether an iris or individual is available or not. Available is made by the performance evaluator.

Unavailable (failure to enroll or acquire) - This term is used to indicate fingerprints that cannot be used in evaluation because there is no available registered or verification data.

Unavailable image - 1) A person who cannot be used in evaluation because, for example, no enrollment data or enrollment-data set cannot be generated for him or her. The evaluator determines whether a person is available or unavailable. 2) An image that cannot be used in evaluation because, for example, part of the face cannot be seen.

Unavailable person - A person who cannot be used in evaluation because, for example, no enrollment data or enrollment-data set cannot be generated for him or her. The evaluator determines whether a person is available or unavailable. .

Unavailable test volunteer - Test volunteers whose voices are unavailable. For example, one of twins, a person who has a voice similar to any other person (wolf and lamb speaker), or a person who has a drastically variable voice (goat speakers).

Uncooperative User - An individual who actively tries to deny the capture of his/her biometric data, Example: A detainee mutilates his/her finger upon capture to prevent the recognition of his/her identity via fingerprint. *See also cooperative user, indifferent user, and non-cooperative user.*

Uniqueness - How well the biometric separates one individual from another.

Universality - Describes how commonly a biometric is found in each individual.

Unknown to the system - A person is "known" to the system if the person is enrolled and the enrollment affects the templates of others in the system.

User - 1) A person who requires access to the portal which is protected by a biometric system. 2) The client to any biometric vendor; a person, such as an administrator, who interacts with or controls end users' interactions with a biometric system; the client to any biometric vendor. The user must be differentiated from the end user and is responsible for managing and implementing the biometric application rather than actually interacting with the biometric system. *See also cooperative user, end user, indifferent user, non-cooperative user, and uncooperative user.*

Usurpator - A person who submits a replication of a legitimate enrollee in either an intentional or inadvertent attempt to pass him/herself as this legitimate enrollee.

Usurpator claim of identity - Making a false positive claim about identity in the system.

V

VAR - Visit Authorization Request

VBIED - Vehicle-Borne Improvised Explosive Device

VBSS – Visit, Board, Search and Seizure

VCJCS - Vice-Chairman, Joint Chiefs of Staff

VCSA - Vice Chief of Staff, Army

VGTOF - Violent Gangs and Terrorist Organization File - Database indexing criminal information and terrorist identities, VGTOF can be made available to police officers on mobile patrol, thereby providing information immediately to frontline law enforcement officials.

VIS - Visa Information System

Validation - The process of demonstrating that the system under consideration meets in all respects the specification of that system.

Valley - The area of a fingerprint surrounding a friction ridge that does not make contact with an incident surface under normal touch.

Veincheck/Vein Tree - A physical biometric that analyses the pattern of veins (e.g., in the back of the hand).

Verification (1:1, matching, authentication) - The process of establishing the validity of a claimed identity by comparing a verification template to an enrollment template. Verification requires that an identity be claimed, after which the individual's enrollment template is located and compared with the verification template. Verification answers the question, "Am I who I claim to be?" Some verification systems perform very limited searches against multiple enrollee records. For example, a user with three enrolled finger-scan templates may be able to place any of the three fingers to verify, and the system performs

1:1 matches against the user's enrolled templates until a match is found. There is a middle ground between identification and verification referred to as one-to-few (1: few). This type of application involves identification of a user from a very small database of enrollees. While there is no exact number that differentiates a 1:*N* from a 1:few system, any system involving a search of more than 500 records is likely to be classified as 1:*N*. A typical use of a 1:few system would be access control to sensitive rooms at a 50-employee company, where users place their finger on a device and are located from a small database; A task where the biometric system attempts to confirm an individual's claimed identity by comparing a submitted sample to one or more previously enrolled templates. *See also identification and watchlist.*

Verification (also) - 1) The "one-to-one" process of comparing a submitted against the biometric reference of a single whose identity is being claimed, to determine whether it matches the enrollee's. 2) The process of using a submitted biometric sample for comparison against a template to match a user to a known enrollee. Normally used only in one-to-one systems where the user may also have to specify a user name and/or password or PIN, in verification systems, the user makes a "positive" claim to an identity, requiring a "one-to-one" comparison of the submitted "sample" biometric measure to the enrolled "template" for the claimed identity. To compare verification data with enrollment data or an enrollment data set, and calculate the matching score between them. Enrollment data on another person may be referred to during matching; the operation of comparing a submitted biometric sample against a specific claimed biometric reference template to determine whether it sufficiently matches that template. *Contrast with identification.*

Verification algorithm - The algorithm for comparing enrollment data with a face image and calculating the degree of matching score. Enrollment data or data sets of another person may be referred to during the calculation of matching score.

Verification attempt - A verification attempt is the presentation of a claimed identity to a biometric identification device together with presentation of a single biometric sample for an identification decision.

Verification data - 1) A record compared with the enrollment data for matching purposes. Normally, the features or face image is used as the verification data. 2) Data compared with enrollment data during matching or matching decision. An attempt creates one item of verification data. Generally, the features or fingerprint images itself are used as verification data. A piece of verification data is generated by a single voice input.

Verification data set - Collection of verification data.

Verification device - 1) The device comprising a matching algorithm and an iris imager, which generates a piece of verification data for each attempt and calculates the matching score or distance by comparing the verification data against a piece of enrollment data. 2) A device consisting of a matching algorithm and fingerprint scanner that creates one item of verification data from one attempt and verifies it against one item of enrollment data to determine matching score or distance

Verification rate - A statistic used to measure biometric performance when operating in the verification task. The rate at which legitimate end-users are correctly verified

Verification requirement information - Indicate whether the reference data for user verification (i.e., passwords and/or biometric data) are enabled or disabled and usable or unusable.

Verification system - Verification systems, where the user explicitly claims an identity, may be compared to identification systems.

Verification unavailability - Verification unavailability indicates that some types of voice data or test volunteers (speakers) cannot be used for verification, for example because the verification-data set cannot be generated.

Voice print/Voiceprint - A representation of the acoustic information found in the voice of a speaker.

Voice recognition - *See speaker recognition.*

Voice verification - *See speaker verification.*

Volatiles - The chemical breakdown of body odor.

Vulnerability - The potential for the function of a biometric system to be compromised by intent (fraudulent activity); design flaw (including usage error); accident; hardware failure; or external environmental condition. *See also threat.*

W

WAN - Wide Area Network

W3C - World Wide Web Consortium

WG - Working Group

WHCA - White House Communications Agency

WHMO - White House Military Office

WHS - Washington Headquarters Service

WHTI - Western Hemisphere Travel Initiative - WHTI requires people traveling to and from Canada, Mexico, the Caribbean, and Bermuda to present a passport or other approved identification document in order to enter or re-enter the United States.

WNINTEL - Warning Notice - Intelligence Sources or Methods Involved (obsolete)

WS - Web Services

WSQ - Wavelet Scalar Quantization - 1) A compression algorithm used to reduce the size of reference templates. 2) An FBI-specified compression standard algorithm that is used for the exchange of fingerprints within the criminal justice community is used to reduce the data size of images.

Watchlist/Watch list - A term sometimes referred to as open-set identification that describes one of the three tasks that biometric systems perform. Answers the questions whether a person is in a given database and, if so, who that person is. The biometric system determines if the individual's biometric template matches a biometric template of someone on the watch list. The individual does not make an identity claim, and in some cases does not personally interact with the system whatsoever. *See also closed-set identification, identification, open-set identification, and verification.*

Wavelet Scalar Quantization (WSQ) Grayscale Fingerprint Image Compression Specification - Provides the definitions, requirements, and guidelines for specifying the FBI's WSQ compression algorithm. Specifies the class of encoders required, decoder process, and coded representations for compressed image data.

Whorl - A fingerprint pattern in which the ridges are circular or nearly circular.

Wiegand - Wiegand is the trade name for a technology used in card readers and sensors, particularly for access control applications. Wiegand devices were originally developed by HID Corporation. A Wiegand card looks like a credit card. It works according to a principle similar to that used in magnetic-stripe cards, such as those used with bank automatic teller machines (ATMs). Instead of a band of ferromagnetic material, the Wiegand card contains a set of embedded wires. The wires are made of a special alloy with magnetic properties that are difficult to duplicate. This makes Wiegand cards virtually counterfeit-proof.

The set of wires can contain data such as credit card numbers, bank account numbers, employee identification information, criminal records, and medical history. The card is read by passing it through, or bringing it near, a device called a Wiegand sensor. Wiegand effect occurs over a wide range of temperatures. Therefore, access control devices using this technology can function in hostile environments. Other assets include rapid response time and portability. These properties make Wiegand cards and readers ideal for use in the field.

X

X.509 Certificate Policy for the U.S. Department of Defense - International standard for security and authentication services supporting security frameworks for electronic information distribution. The term "X.509 certificate" has become the *de facto* name for public key certificates in use today. This specification defines the main data structure (i.e., the "certificate") used for performing these services and addressing the handling of keys.

X9.84 Biometrics Management and Security for the Financial Services Industry - Specification that defines the minimum-security requirements for effective management of biometrics data for the financial services industry and the security for the collection, distribution and processing of biometrics data.

XDI - XRI Data Interchange

XRI - eXtensible Resource Identifier

Y

Y STR analysis - Type of nuclear DNA analysis involving the Y sex chromosome, typically found only in males. The Y chromosome is inherited paternally. Therefore, all males of a specific paternal line still have the same Y STR DNA profile. Y STR analysis is used to support kinship analyses.

Z

Zero effort forgery - Where an impostor uses his or her own biometric sample and claims the identity of a different enrollee.

Zero-effort attempts - An impostor attempt is classed as "zero-effort" if the individual submits their own biometric feature as if they were attempting successful verification against their own template.

Sources

Abstract Syntax Notation One (http://asn1.elibel.tm.fr)

Accenture (www.accenture-blogpodium.nl/veiligheid/
gideon-shimshon/biometrics-terminology-a-short-guide)

Acquisition Community Connection (https://acc.dau.mil/
CommunityBrowser.aspx?id=22231)

Advanced Systems and Concepts website (www.acq.osd.
mil/asc)

American Association of Motor Vehicle Administrators
(AAMVA) National Standards for the Driver License/
Identification Card

American National Standards Institute/National Institute for
Standards and Technology (ANSI/NIST-ITL 1-2000), Data
Format for the Interchange of Fingerprint, Facial, & Scar
mark & Tattoo Information

American National Standards Institute/InterNational
Committee for Information Technology Standards (ANSI/
INCITS) 358-2002, Information Technology: BioAPI
Specification, (16 March 2001)

American National Standards Institute (ANSI)/X9 X9.84-
2001, Biometric Information Management and Security, (27
March 2001)

Richard D. Newbold

Army Knowledge Online (www.us.army.mil)

Association for Biometrics (AfB) and International
Computer Security Association (ICSA), Glossary of
Biometric Terms, (1999)

AuthenTec website (http://www.authentec.com/technology-
glossary.html)

Bankrate (www.bankrate.com)

BioID Glossary

Biometrics: A Grand Challenge, Proceedings of
International Conference on Pattern Recognition,
Cambridge, UK, (August 2004)

Biometrics Technologies, Biometric Glossary

Categorization of Federal information and Information
Systems, Center for Identification Technology Research
(www.citer.wvu.edu), (February 2004)

Center for Identification Technology Research (CITeR)
(www.citer.wvu.edu/about/index.php)

Committee on National Security Systems, CNSS
Instruction No. 4009, National Information Assurance (IA)
Glossary (May 2003)

Common Criteria for Information Technology Security
Evaluation, Version 2.1, Parts 1, 2, and 3 (August 1999)

Cornell Law School (www4.law.cornell.edu/uscode/search/
display.html?terms=403&url=/uscode/html/uscode41/usc_
sec_41_00000403----000-.html)

Criminal Justice Information Services (CJIS) Electronic
Fingerprint Transmission Specification

Defence Diversification Agency (www.dda.gov.uk)

Defense Biometric Identification System (www.dmdc.osd.
mil/iao/pages/dbids/dbids_main.html)

Defense Information Systems Agency (DISA) website,
Public Key Infrastructure Terms and Definitions

Defense Intelligence Agency (DIA) (www.dia.mil)

Defense Manpower Data Center (DMDC) (www.dmdc.osd.
mil/)

Defense Security Service (DSS) (www.dss.mil/about_dss/
fact_sheets/dss_faqsheet.html)

Department of Defense (DoD) (www.dod.mil/odam/index.
htm)

Department of Defense (DoD) Biometrics Fusion Center
(BFC)/Biometrics Task Force (BTF) (www.biometrics.dod.
mil)

Department of Defense (DoD) Chief Information Officer
(CIO) Memorandum, Department of Defense (DoD) Public
Key Infrastructure (PKI), (August 2000)

Department of Defense (DoD) Deputy Secretary of Defense (DEPSECDEF) Memorandum, Smart Card Adoption and Implementation, (10 November 1999)

Department of Defense Directive (DoDD) 3224.3, (17 February 1989)

Department of Defense Directive (DoDD) 4640.6, Communications Security Telephone Monitoring and Recording, (26 June 1981)

Department of Defense (DoD) Public Key Infrastructure (PKI) and Key Management Infrastructure Token Protection Profile (Medium Robustness), Common Criteria for Information Technology Security Evaluation

Department of Defense (DoD) Public Key Infrastructure (PKI) Program Management Office, X.509 Certificate Policy for the United States Department of Defense (18 December 2003)

Department of Justice (DoJ) (www.usdoj.gov/oip/privstat.htm)

Department of Justice (DoJ)/Federal Bureau of Investigation (FBI) Technical Specifications Document for the Repository for Individuals of Special Concern (RISC) Prototype Project Version 1.0 (8 June, 2007)

Entrust Resources Security Glossary

Eyenetwatch.com (www.eyenetwatch.com/biometrics-glossary/m.htm)

Federal Bureau of Investigation (FBI) (www.fbi.gov)

Federal Emergency Management Agency (FEMA) Information Technology Architecture, The Road to FEMA

Federal Information Processing Standards Publication (FIPS) 199, Standards for Security

Federation of American Scientists (FAS) (www.fas.org/ nuke/guide/usa/c3i/raven_rock.htm)

Federation for Identity and Cross-Credentialing Systems (FiXS) (www.fixs.org)

Find Biometrics (http://www.findbiometrics.com/Pages/ glossary.html)

Foreseeable Risk Analysis Center (FRAC) (www.frac.com)

FreePatentsOnline (www.freepatentsonline.com)

Future Combat System website (www.army.mil/fcs/)

GlobalSecurity (www.globalsecurity.org)

Global Village, Technologies Glossary

Headquarters, Department of the Army website (www. hqda.army.mil)

HRS Technologies, LLC website (http://www. hrstechnologiesllc.com/HRSTLLC/HTML/glossary.htm)

Richard D. Newbold

Immigration and Customs Enforcement (ICE) (http://www. ice.gov)Naval Criminal Investigative Service (NCIS) (http:// www.ncis.navy.mil/mission/mtac.asp)

Information Systems (Second Public Draft), (April 2006)

Intel Corporation website, Biometric User Authentication: Fingerprint Sensor Product Guidelines Version 1.03, (September 2003)

International Association for Biometrics (www.afb.org.uk/ docs/glossary.htm)

International Organization of Standardization (ISO)/ International Electrotechnical Commission (IEC) TR 10000-1:1998 (E) Information Technology-Framework and Taxonomy of International Standardized Profiles

IQBio (http://www.iqbio.com/en_us/content/biometricterms. htm)

J2 Intelligence Directorate (www.jfcom.mil/about/abt_ j2.htm)

Joint Improvised Explosive Device Defeat Organization (JIEDDO) (https://www.jieddo.dod.mil/)

Joint Publication (JP) 1-02, Department of Defense (DoD) Dictionary of Military and Associated Terms (17 October 2007)

National Biometric Test Center Collected Works (1997-2000), Version 1.2, (August 2000)

National Consumers League (NCL) Advocacy, CDT Working Group on RFID, Privacy Best Practices for Deployment of RFID

National Counterterrorism Center (www.nctc.gov)

National Defense Industrial Association website (www.ndia.org)

National Information Assurance Partnership, Common Criteria Evaluation Validated Scheme

National Information Assurance Partnership, US Government Biometric Verification Mode Protection Profile for Medium Robustness Environments, v1.0, Sponsored by the DoD Biometrics Management Office (BMO) and the National Security Agency (NSA), (15 November 2003)

National Institute of Standards and Technology (http://iris.nist.gov/ICE)

National Institute of Standards and Technology Interagency Report (NISTIR) 6529-2001, Common Biometric Exchange File Format, (3 January 2001)

National Institute of Standards and Technology (NIST) Special Publication 800-53, Recommended Security Controls for Federal Information Systems, (February 2005)

National Institute of Standards and Technology (NIST) Special Publication 800-53A, Guide for Assessing the Security Controls in Federal

Richard D. Newbold

National Law Enforcement and Corrections Technology
Centre (http://www.nlectc.org)

National Security Agency, Guidelines for Placing
Biometrics in Smartcards, Draft v1.0, 15 (September 1998)

Naval Facilities Engineering Service Center, Antiterrorism
Team, Glossary of Terms

North Atlantic Treaty Organization (NATO) Glossary of
Terms and Definitions AAP-6 (www.nato.int/docu/stanag/
aap006/aap6.htm)

Office of the Assistant Secretary of Defense (Command,
Control, Communications & Intelligence), Automated
Document Conversion Master Plan, (April 1995)

Office of the Assistant Secretary of Defense (Command,
Control, Communications & Intelligence), DoD Web Site
Administration Policies & Procedures, (11 January 2002)

Office of the Secretary of Defense (OSD) Comparative
Testing Office (CTO) (https://cto.acqcenter.com/osd/portal.
nsf)

Office of the Under Secretary of Defense for Acquisition,
Technology and Logistics (www.acq.osd.mil/dsb)

Pentagon Area Common Information Technology (IT)
Wireless Security Policy, (September 2002)

Practitioner's Guide to Biometrics (American Bar
Association, 2007)

Privacy Technology Implementation Guide, Department of Homeland Security (August 16, 2007)

Report of the Defense Science Board Task Force on Defense Biometrics (March 2007)

Scientific Working Group on Friction Ridge Analysis, Study, and Technology (SWGFAST) (www.swgfast.org)

SecuriMetrics (www.securimetrics.com)

Symmetricom, Information Center

Technology Interim Draft (1 May, 2006)

Transportation Security Administration (TSA) http://www.tsa.gov/what_we_do/layers/twic/index.shtm

Under Secretary of Defense for Personnel and Readiness (www.defenselink.mil/prhome)

United States Army Aeromedical Research Laboratory website, CAC/ PKI Definitions and FAQs

Webopedia (www.webopedia.com)

Wikipedia (www.wikipedia.org)

E-mail the author at
richard.d.newbold@gmail.com.